An Introduction to Mathematical Ecology

An Introduction to Mathematical Ecology

E. C. PIELOU
Professor of Mathematical Biology
Queen's University, Kingston, Ontario

WILEY-INTERSCIENCE A Division of John Wiley & Sons
New York · London · Sydney · Toronto

10 9 8 7 6 5 4 3 2 1

Library of Congress Catalogue Card Number: 77-88610

SBN 471 68918 1

Printed in the United States of America

Preface

The fact that ecology is essentially a mathematical subject is becoming ever more widely accepted. Ecologists everywhere are attempting to formulate and solve their problems by mathematical reasoning, using whatever mathematical knowledge they have acquired, usually in undergraduate courses or private study. The purpose of this book is to serve as a text for these students and to demonstrate the wide array of ecological problems that invite continued investigation.

In writing a book of this length (or, indeed, of any length) the author is always faced with the problem whether to cover a great many topics superficially or to delve deeply into a few. The compromise I have attempted has been to pick selected topics over the whole range of mathematical ecology and then to deal in detail with those aspects that seem likely to furnish good starting points for further research. The list of chapter headings shows the subjects that were chosen. Their choice, and the aspects to be pursued, has, of course, been subjective and a matter of my own judgment. It is unlikely that any other ecologist with the same objective would have selected exactly the same topics.

The book is in no sense a review. An enormous amount of interesting and valuable work has been ignored and any mathematical ecologist glancing at the bibliography is bound to wonder at what may seem to him inexplicable omissions. The decision to exclude an account of an interesting piece of work was always difficult but the exclusions were necessary to make room for a fairly full exposition of the matters dealt with. My object has been to provide a sufficiently detailed development of the topics examined for the reader to be able to consult the current literature with an understanding of what has led up to it; as far as possible nothing has been asserted without proof. For each topic I have attempted to give a connected account of the underlying theory. So that the reader will not lose the thread of an argument, methods of estimating parameters and similar practical details have not been given. Information on these matters, and numerical examples, will be found in the literature cited.

To make the text suitable for private study, many mathematical derivations have been written out *in extenso*. I have striven to avoid the mathematician's daunting phrases "Obviously therefore . . ." and "Clearly then . . ." which are so often substituted for long chains of reasoning that, for the mathematically inexperienced, are neither obvious nor clear. When these phrases are used here, a line or two at most of algebraic manipulation is all that the reader needs to interpolate.

This book is based on a one-semester graduate course in mathematical ecology that I gave in the spring of 1968 while I was Visiting Professor in the Biomathematics Program of the Department of Experimental Statistics, North Carolina State University, Raleigh. It is a pleasure to thank Dr. H. L. Lucas Jr., Director of the Program, for his hospitality.

Above all, I thank my husband D. P. Pielou for his constant help and encouragement.

E. C. Pielou

Kingston, Ontario
July, 1969

Contents

An Introduction to Mathematical Ecology

Introduction

Most ecological communities are made up of a vast profusion of living things. In an acre of forest, for instance, an enormous number of species is present, from trees to soil microorganisms. Not only does each species differ from every other, but also all the individuals within any one species are unique. Each individual is a complex organism that changes continuously; at any moment its behavior depends on its genetic constitution, its age, the vicissitudes of its life up to that moment, and on the conditions that prevail in its locality at that moment. Influences too numerous to list affect its physiological processes unceasingly. It can truly be said that no two of the individual units making up a community are alike and that each of them, throughout its lifetime, varies continuously in a manner peculiar to itself. In addition to all this, a community as a whole is at all times being depleted by deaths and replenished by births; often immigration and emigration are further causes of a continual turnover of individuals. Thus the components of a community are never the same on two successive occasions.

In spite of all this unending change, however, experience shows that in the absence of outside disturbance most communities either remain in a steady state for long periods or pass through an orderly progression of successional stages culminating in a steady state. The concept of a steady state is, of course, a subjective notion. It seems reasonable to say, however, that a steady state exists if the gross properties of a community persist for periods many times as long as the lifetimes of most of the species present. This cautious definition allows for the possibility that the longest lived species in a community may never attain populational stability, and if apparent stability prevails this is merely because of the comparatively short lifetimes of human observers (see Frank, 1968). A community undergoing cyclical fluctuations is also in a steady state, of course, provided the cycles are regular and no trend is superimposed on them.

In any case, that steady states should be as steady and successions as orderly, as they are found to be, is remarkable when one considers the unceasing internal changes going on in a community. To say that most communities persist for generation after generation of all but a few of their constituent species is, indeed, a tautology, for if they did not do so there would be no recognizable entities that could be called communities. Although probably no two ecologists hold identical opinions over the degree to which assemblages of organisms can be regarded as autonomous and self-contained, there is no need to insist that an assemblage should have these properties for it to be defined as a community. Anyone who admits the existence of such obvious features of a landscape as forests, meadows, and swamps, realizes that the distinctness and persistence of these things deserve to be explained.

In contrast to the steadiness stressed above are the spectacular population explosions that occasionally occur. A single species, which may be a regular member of a community or a chance immigrant, rapidly increases in numbers until it has brought about profound, and sometimes irreversible, change in its surroundings. If the change is deleterious from a human point of view, the species that explodes is by definition a pest, and the latest addition to the ever-growing list of injurious pests is, of course, man himself. (Anyone who doubts this has only to ask himself: which pest is responsible for the serious modern problem of environmental pollution?) Pest control consists in preventing the population of any species from rising to a level at which it does appreciable economic damage and is one of the most important branches of applied ecology. If what is being damaged is a natural community (as opposed to a cultivated crop, stored products, or a human artefact) pest control thus consists in ensuring that the community remains in what is believed to be its natural steady state.

In very general terms, then, there are two classes of questions confronting ecologists. In the first place, what are the processes that permit maintenance of a steady state, or the gradual, orderly succession of states, that occurs in "healthy" undisturbed communities? Second, what are the causes and consequences of sudden departures from steadiness?

The development of theories to account for both these things has come to be known as ecological model building. The investigator tries to envisage a hypothetical system that is sufficiently realistic in the biological sense to be at least approximately true and that by simulating natural processes enables him to predict future developments in the real system with at least approximate correctness. Thus a successful model is one that explains what is currently happening and, if change is expected, predicts what will happen.

At present there are two schools of thought on the subject of model building. The contrast between them has been well described by Leigh (1968). The members of one school insist that models be simple and take pains to keep the assumptions to a minimum, even though the true state of affairs in the real world is almost certainly not simple. By postulating a simple model one may explore its full consequences by mathematical reasoning.

The other school has come into being since high-speed computers became readily available. Its disciples need no longer limit themselves to models that permit elegant mathematical analyses; they do not seek for solutions capable of concise symbolic expression and for equations that can be solved on a desk calculator. Instead, they are free to make models as complicated as they wish and let a computer work out the consequences. It may be argued that a simple model can never be realistic, hence that realism demands complicated models. But the objection to complicated models is that as soon as one foresakes the Occam's razor principle* he runs a risk of constructing a model that incorporates more postulates than are strictly necessary; if unrestricted proliferation of the postulates is permitted, the number of plausible models that will simulate any given sequence of events becomes conceptually infinite and there is no criterion for choosing among them.

At the moment theoretical ecologists seem to be about evenly divided between the makers of simple models and the makers of complicated ones. In this book we describe simple, mathematically tractable models. For a discussion of complicated model building the reader is referred to Watt (1968).

The distinction between simple and complicated models should not be confused with the distinction between deterministic and stochastic (or probabilistic) ones. Model builders of both schools described above concede the need for stochastic models whenever possible. Whether or not one regards all natural occurrences as fundamentally determinate, it is clearly impracticable to treat them as such, and a model in which chance mechanisms are incorporated is sure to be nearer the truth than a deterministic one. Even so, some of the classical deterministic models of early demographers are still worth examining as a preliminary to studying their modern stochastic versions.

Although the greater part of this book is concerned with ecological models, the last four chapters are devoted to a consideration of some of the ways in which ecological field data can be simplified and clarified. There has been much recent work on the application of multivariate statistical methods in ecological contexts and it is yielding interesting results; the present aim of these studies is to describe, rather than to explain, the structure of natural communities. However, until they are described in a way that can be grasped attempts at explanation are futile.

* The maxim is usually stated as "Entities are not to be multiplied without necessity," but, according to Russell, what William of Occam (*ca* 1290–*ca* 1350) actually said was "It is vain to do with more what can be done with fewer." In either version the gist is clear (see B. Russell: *A History of Western Philosophy*, Geo. Allen and Unwin, 1947).

An introduction necessarily deals with broad generalities. One should not, however, draw the conclusion that mathematical ecology, as a subject, is a unified whole; far from it. Anyone who is dismayed by the seemingly fragmentary nature of the work described in this book must console himself with the thought that the subject is still in its early stages and that the welding together of its disconnected parts is a challenging job yet to be done.

I

Population Dynamics

1

Birth and Death Processes

1. Introduction

All populations of organisms fluctuate in size. For any population the only assertion that can be made about it with certainty is that its size will not remain constant. The investigation of the growth and decline of populations is, historically, the oldest branch of mathematical ecology. It has been found that a fruitful way to proceed is to postulate a simple model to account for population change and then to examine the consequences of the assumptions by mathematical argument. We therefore begin by considering the simplest of all systems, the so-called pure birth process, and then deal with the simple birth and death process.

Before doing so, however, it is worthwhile to justify this approach. Field ecologists are often understandably impatient with arguments that proceed from assumptions so greatly oversimplified as to be manifestly unreasonable. But much can be said in defence of the simple, abstract models. Obviously one must study the behavior of simple models before modifying and complicating the first, simplest assumptions, and the simple models provide a basis for elaboration. Also, the way in which natural populations differ from these simple models in their behavior may suggest in what way the over-simplified assumptions are false and how they should be altered. Finally, if models are required for *ad hoc* predictive purposes, simple models are often adequate, even though we can feel no confidence that the underlying model gives a true explanation of the natural process being modeled. This is especially true when we are concerned with population changes that occur over short periods of time. The resemblance between a natural population and a simple model may then be very close.

2. The Pure Birth Process

In the pure birth process the assumptions are as follows: The organisms are assumed to be immortal and to reproduce at a rate that is the same for every individual and does not change with time. It is also assumed that the individuals have no effect on one another. Since no deaths occur, a population growing in this manner can only increase or remain constant; it cannot decrease. In spite of the extreme simplicity of these assumptions they might apply, approximately at least and over a short time interval, to the growth of a population of single-celled organisms that reproduce by dividing. Possibly algal blooms in eutrophic lakes increase in this manner in spring.

If we write N_t for the size of the population at time t and λ for the rate of increase of each individual, it is clear that

$$\frac{dN_t}{dt} = \lambda N_t,$$

hence that $\ln N_t = \lambda t + C$, where C is a constant of integration. Suppose that the initial size of the population at time $t = 0$ were i. Then C may be evaluated, since at $t = 0$, $\ln i = C$. Therefore

$$\ln \frac{N_t}{i} = \lambda t \quad \text{or} \quad N_t = ie^{\lambda t}.$$

This is the Malthusian equation for population growth and it shows that such growth is exponential in the simple circumstances postulated.

The process just described is, of course, *deterministic*. It assumes not that an organism may reproduce but that in fact it does reproduce with absolute certainty. Clearly, however, population growth is a stochastic process. Given a population of yeast cells growing by fission, for example, one can say only that there is a certain probability that a particular cell will divide in a given time interval. We must therefore investigate the stochastic form of the pure birth process.

Suppose, then, that in any short time interval Δt the probability that a cell will divide is $\lambda\, \Delta t + o(\Delta t)$. Here $o(\Delta t)$ denotes a quantity of smaller order of magnitude than Δt. Thus the probability of a birth in a population of size N is $\lambda N\, \Delta t$, plus terms of smaller order of magnitude than Δt. Then the probability that the population is of size N at time $t + \Delta t$ is

$$p_N(t + \Delta t) = p_{N-1}(t)\, \lambda(N - 1)\, \Delta t + p_N(t)(1 - \lambda N\, \Delta t);$$

that is, either the population was of size $N - 1$, with probability $p_{N-1}(t)$ at time t, and a division occurred in Δt or the population was of size N at time t and no division occurred in Δt. Then

$$\frac{p_N(t + \Delta t) - p_N(t)}{\Delta t} = -\lambda N p_N(t) + \lambda(N - 1)p_{N-1}(t).$$

Allowing Δt to tend to zero and putting

$$\lim_{\Delta t \to 0} \frac{p_N(t+\Delta t) - p_N(t)}{\Delta t} = \frac{d\,p_N(t)}{dt},$$

we see that

$$\frac{dp_N(t)}{dt} = -\lambda N p_N(t) + \lambda(N-1)p_{N-1}(t). \tag{1.1}$$

Starting from this differential difference equation, we now wish to find $p_N(t)$ in terms of N, λ, t and the initial size of the population at time $t=0$, again denoted by i; that is $p_i(0)=1$ and $p_N(0)=0$ when $N \neq i$. It is also clear that since no deaths occur and consequently the population size can never be less than its initial value we must have $p_{i-1}(t)=0$. From these conditions it follows from (1.1) that

$$\frac{dp_i(t)}{dt} = -\lambda i p_i(t). \tag{1.2}$$

We shall now solve (1.2) to obtain $p_i(t)$, substitute the result in (1.1), solve for $p_{i+1}(t)$, and so on. In this manner $p_N(t)$ for all $N > i$ may be found. From (1.2) it is seen that

$$\ln p_i(t) = -\lambda i t + \text{const}.$$

Since, when $t=0$, $p_i(t)=1$ and $\ln p_i(t)=0$, the constant is also 0 and therefore

$$p_i(t) = e^{-\lambda i t}.$$

This is the probability that during time t no reproduction will occur. To find $p_{i+1}(t)$ we now solve (1.1) in the form

$$\frac{dp_{i+1}(t)}{dt} + \lambda(i+1)p_{i+1}(t) = \lambda i p_i(t) = \lambda i e^{-\lambda i t}.$$

Now multiply both sides by $e^{\lambda(i+1)t}$ so that the left-hand side may be integrated

$$e^{\lambda(i+1)t}\left\{\frac{dp_{i+1}(t)}{dt} + \lambda(i+1)p_{i+1}(t)\right\} = \lambda i e^{\lambda t}.$$

Then

$$e^{\lambda(i+1)t}p_{i+1}(t) = ie^{\lambda t} + \text{const}.$$

Since $p_{i+1}(0)=0$, the constant is $-i$, whence

$$p_{i+1}(t) = ie^{-\lambda i t}(1 - e^{-\lambda t}).$$

Substituting this result in (1.1) so that the equation may be solved for $p_{i+2}(t)$, we have

$$\frac{dp_{i+2}(t)}{dt} + \lambda(i+2)p_{i+2}(t) = \lambda(i+1)ie^{-\lambda it}(1 - e^{-\lambda t}).$$

Multiplying both sides by $e^{\lambda(i+2)t}$ and integrating gives

$$\begin{aligned} e^{\lambda(i+2)t}p_{i+2}(t) &= (i+1)i \int \lambda e^{2\lambda t}(1 - e^{-\lambda t})\,dt \\ &= (i+1)i\,\frac{(e^{\lambda t} - 1)^2}{2} + \text{const.} \end{aligned}$$

Since $p_{i+2}(0) = 0$, const. $= 0$ and

$$p_{i+2}(t) = \frac{(i+1)i}{2}\,e^{-\lambda it}(1 - e^{-\lambda t})^2.$$

The form of the general solution now becomes evident. It is

$$p_N(t) = \binom{N-1}{i-1} e^{-\lambda it}(1 - e^{-\lambda t})^{N-i} \tag{1.3}$$

and its correctness may be proved by induction. Assuming (1.3) to be true, $p_{N+1}(t)$ may be found by solving

$$\frac{dp_{N+1}(t)}{dt} + \lambda(N+1)p_{N+1}(t) = \lambda N \binom{N-1}{i-1} e^{-\lambda it}(1 - e^{-\lambda t})^{N-i}$$

or

$$\begin{aligned} e^{\lambda(N+1)t}p_{N+1}(t) &= N\binom{N-1}{i-1}\int \lambda e^{\lambda t}(e^{\lambda t} - 1)^{N-i}\,dt \\ &= N\binom{N-1}{i-1}\frac{(e^{\lambda t} - 1)^{N-i+1}}{N-i+1} + \text{const.} \end{aligned}$$

The constant is 0, since $p_{N+1}(0) = 0$, and therefore

$$\begin{aligned} p_{N+1}(t) &= \frac{N}{N-i+1}\binom{N-1}{i-1} e^{-\lambda(N+1)t}(e^{\lambda t} - 1)^{N-i+1} \\ &= \binom{N}{i-1} e^{-\lambda it}(1 - e^{-\lambda t})^{N-i+1}, \end{aligned}$$

which is of the same form as (1.3).

Notice that in the formula for $p_N(t)$, λ and t cannot be separated; they occur only in the form of the product λt. It follows that the probability distribution of N, when i is given, depends only on λt. Thus a high reproductive rate acting over a short time will yield the same result as a low rate over a longer time if λt is the same for both. Figure 1 shows an example of the distribution when $\lambda t = 0.5$ and $i = 5$. The mean and variance of the distribution may be derived as follows.

Write $M(N \mid t)$ for the expected size of the population at time t; that is,

$$\begin{aligned} M(N \mid t) &= \sum_{j=i}^{\infty} j p_j(t) \\ &= e^{-\lambda i t} \sum_{k=0}^{\infty} (i+k) \binom{i+k-1}{k} (1-e^{-\lambda t})^k \\ &= i e^{-\lambda i t} \sum_{k=0}^{\infty} \binom{i+k}{k} (1-e^{-\lambda t})^k \\ &= i e^{-\lambda i t} (e^{-\lambda t})^{-(i+1)} \\ &= i e^{\lambda t}. \end{aligned}$$

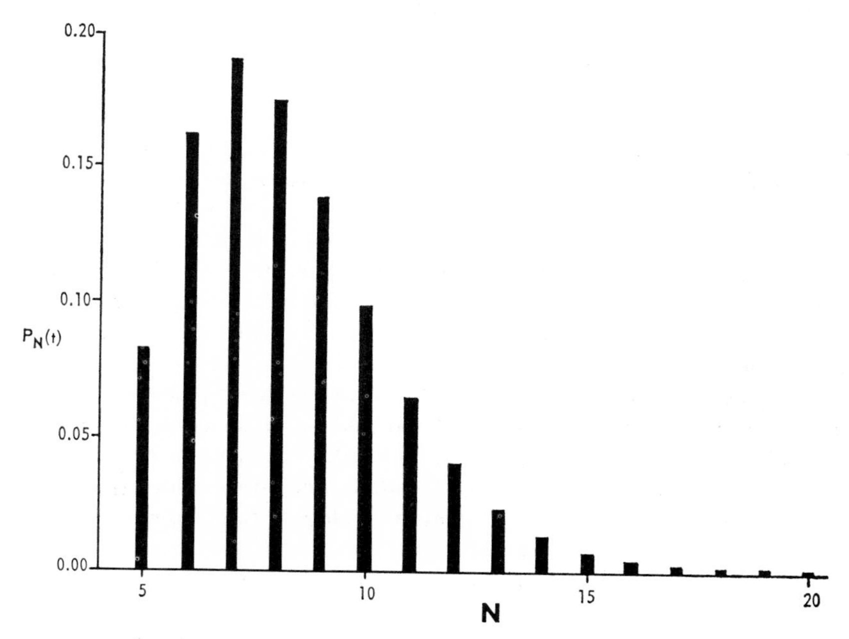

Figure 1. The probability distribution of the size at time t of a population undergoing a pure birth process:

$$p_N(t) = \binom{N-1}{i-1} e^{-\lambda i t} (1 - e^{-\lambda t})^{N-i},$$

with $\lambda t = 0.5$ and $i = 5$. The mean is $M(N \mid t) = i e^{\lambda t} = 8.24$. The variance is $\text{var}(N \mid t) = i e^{\lambda t} (e^{\lambda t} - 1) = 5.35$.

Thus the expected population size in the stochastic case is equal to that predicted with certainty for the deterministic process.

Now write $M_2(N \mid t)$ for the second moment about the origin; that is,

$$
\begin{aligned}
M_2(N \mid t) &= \sum_{j=i}^{\infty} j^2 p_j(t) \\
&= e^{-\lambda it} \sum_{k=0}^{\infty} (i+k)^2 \binom{i+k-1}{k} (1-e^{-\lambda t})^k \\
&= i^2\, e^{-\lambda it} \sum_{k=0}^{\infty} \binom{i+k}{k} (1-e^{-\lambda t})^k + ie^{-\lambda it} \sum_{k=0}^{\infty} k \binom{i+k}{k} (1-e^{-\lambda t})^k \\
&= i^2\, e^{\lambda t} + (i+1)ie^{-\lambda it}(1-e^{-\lambda t}) \sum_{k=0}^{\infty} \binom{i+k+1}{k} (1-e^{-\lambda t})^k \\
&= i^2\, e^{\lambda t} + (i+1)i(e^{\lambda t}-1)\, e^{\lambda t}.
\end{aligned}
$$

Then the variance $\text{var}(N \mid t)$ is given by

$$
\begin{aligned}
\text{var}(N \mid t) &= M_2(N \mid t) - [M(N \mid t)]^2 \\
&= ie^{\lambda t}(e^{\lambda t}-1).
\end{aligned}
$$

As we would expect, the variance increases with t. This amounts to saying that the farther into the future we make predictions the less precise these predictions will be.

The process described by (1.1) is known as the Yule process and was first proposed by Yule (1924) in connection with evolution to describe the rate of evolution of new species within a genus. Because of the very stringent and, in most circumstances, unrealistic assumptions on which it is based, it is of limited usefulness in ecology.

The opposite process, the pure death process, is likely to be more widely applicable. In this process it is postulated that no births occur and that for each individual in the population the probability of dying in an interval of length Δt is constant and equal to $\mu\, \Delta t + o(\Delta t)$. The rate at which the population declines is then given by an equation analogous to (1.1), namely

$$
\frac{dp_N(t)}{dt} = \mu(N+1)p_{N+1}(t) - N\mu p_N(t).
$$

By arguments similar to those already given for the pure birth process we may now find the probability $p_N(t)$ that at time t the population is of size N, given that its initial size was i, $(i > N)$.

It will be found that

$$
p_N(t) = \binom{i}{N} e^{-\mu it}(e^{\mu t}-1)^{i-N}. \tag{1.4}
$$

The mean and variance of this distribution are

$$M(N \mid t) = ie^{-\mu t}$$

and

$$\mathrm{var}(N \mid t) = ie^{-\mu t}(1 - e^{-\mu t}).$$

An alternative way of deriving $p_N(t)$ is as follows: Since the probability that an individual will die in any interval Δt is $\mu\,\Delta t + o(\Delta t)$, the probability that it will be a survivor at time t is $e^{-\mu t}$ and the probability that it will die on or before t is $1 - e^{-\mu t}$. (This may be proved by the method given on page 9, where the probability of no births in time t, given a pure birth process, was derived; an alternative proof is given on page 25.) The fate of the i individuals present at $t = 0$ may be thought of as the outcome of i Bernoulli trials for which the probability of "success" (=survival) is $e^{-\mu t}$ and the probability of "failure" (=death) is $1 - e^{-\mu t}$. Then $p_N(t)$ is the probability of N successes in these trials and consequently

$$p_N(t) = \binom{i}{N}(e^{-\mu t})^N(1 - e^{-\mu t})^{i-N},$$

which is the same as (1.4). The mean and variance follow immediately.

We may also find the expectation of the time to extinction. It is i/μ; that is, we expect the last survivor of a pure death process to die at time i/μ, given that the population was of size i at $t = 0$. The way in which this result is derived is mentioned in Chapter 10.

Natural circumstances in which a pure death process will occur seem quite likely to develop. If the environment of an isolated population is altered, by pollution, for instance, in such a way that all reproduction is prevented and if also the death rate of population members is independent of their ages, a pure death process will ensue. It is known (Deevy, 1947) that for many species of birds the death rate is nearly independent of age once they have become adult, and the same is possibly true of many fish.

3. The Simple Birth and Death Process

So far we have assumed that either births or deaths occur exclusively. Now suppose that both births and deaths take place and, as before, assume that an individual's probability of giving birth or of dying is independent of its age and of the size of the population to which it belongs. We also suppose that all members of the population are capable of reproducing; either the organisms are asexual or, if the species is bisexual, we consider only the females and postulate that there is never a shortage of males.

Writing λ for the birth rate and μ for the death rate of each individual, we obtain the deterministic equation for population growth:

$$N_t = ie^{(\lambda-\mu)t}.$$

This is formally the same as the equation for the deterministic form of the pure birth process (see page 8) but here the rate of increase $\lambda - \mu$ is the difference between a birth rate and a death rate; it may be positive or negative.

Presumably natural populations may sometimes grow exponentially in this manner, as long as the numbers are low enough for competition among population members to have no effect on the rate of increase. The growth of such a population may be treated as a simple birth and death process even when the birth and death rates are *not* independent of age, provided the population has a stable age distribution (see Chapters 3 and 4). Then, although the rates differ from one individual to another, the *mean* rate per individual remains constant because the proportions of the population in each age class do not change.

Now consider the stochastic form of the simple birth and death process. In a population of size N, and during an interval of length Δt, let the probability of a birth be $N\lambda\,\Delta t + o(\Delta t)$ and of a death be $N\mu\,\Delta t + o(\Delta t)$. The probability that more than one event (birth or death) will occur in Δt is treated as negligible. Thus, if the population is of size N at $t + \Delta t$, its size at time t must have been $N - 1$, N, or $N + 1$ and we see that

$$p_N(t + \Delta t) = p_{N-1}(t)\,\lambda(N-1)\,\Delta t + p_N(t)[1 - N\lambda\,\Delta t - N\mu\,\Delta t] + p_{N+1}(t)\,\mu(N+1)\,\Delta t.$$

Therefore

$$\frac{dp_N(t)}{dt} = -N(\lambda+\mu)\,p_N(t) + \lambda(N-1)p_{N-1}(t) + \mu(N+1)\,p_{N+1}(t). \quad (1.5)$$

From this equation we now determine the mean and variance of the population size at time t directly; that is, without first deriving the general term $p_N(t)$ of the distribution. (The same method could have been used in connection with the pure birth process.)

As before, write $M(N \mid t)$ for the expected size of the population at time t; that is

$$M(N \mid t) = \sum_{j=1}^{\infty} j\,p_j(t).$$

Then

$$\frac{dM(N \mid t)}{dt} = \sum_{j=1}^{\infty} j\,\frac{dp_j(t)}{dt}.$$

Putting p_j in place of $p_j(t)$ for brevity and substituting from (1.5), we then see that

$$\begin{aligned}\frac{dM(N \mid t)}{dt} &= \sum_{j=1}^{\infty} [-(\lambda+\mu)j^2 p_j + \lambda j(j-1)p_{j-1} + \mu j(j+1)p_{j+1}] \\ &= \lambda \sum_{j=1}^{\infty} (-j^2 p_j + j^2 p_{j-1} - j p_{j-1}) - \mu \sum_{j=1}^{\infty} (j^2 p_j - j^2 p_{j+1} - j p_{j+1}) \\ &= \lambda \sum_{k=0}^{\infty} p_k[-k^2 + (k+1)^2 - (k+1)] \\ &\qquad - \mu \sum_{k=1}^{\infty} p_k[k^2 - (k-1)^2 - (k-1)] \\ &= \lambda \sum_{k=0}^{\infty} k p_k - \mu \sum_{k=1}^{\infty} k p_k = (\lambda - \mu) M(N \mid t).\end{aligned}$$

We now have a differential equation in $M(N \mid t)$ that is easily solved. Thus

$$\frac{dM(N \mid t)}{M(N \mid t)} = (\lambda - \mu)\, dt,$$

whence $M(N \mid t) = e^{(\lambda-\mu)t} \times \text{const.}$ Since at $t = 0$ the population was of size i, that is, $M(N \mid 0) = i$, we then get $M(N \mid t) = ie^{(\lambda-\mu)t}$.

Again, as was true of the pure birth process, the expected population size at time t is identical with that predicted for the deterministic process. Writing $M_2(N \mid t)$ for the second moment about the origin,

$$M_2(N \mid t) = \sum_{j=1}^{\infty} j^2\, p_j(t) \quad \text{and} \quad \frac{dM_2(N|t)}{dt} = \sum_{j=1}^{\infty} j^2 \frac{dp_j(t)}{dt}.$$

Then, from (1.5),

$$\begin{aligned}\frac{dM_2(N \mid t)}{dt} &= \lambda \sum_{j=1}^{\infty} (-j^3 p_j + j^3 p_{j-1} - j^2 p_{j-1}) - \mu \sum_{j=1}^{\infty} (j^3 p_j - j^3 p_{j+1} - j^2 p_{j+1}) \\ &= \lambda \sum_{k=0}^{\infty} p_k(2k^2 + k) - \mu \sum_{k=1}^{\infty} p_k(2k^2 - k) \\ &= 2(\lambda - \mu) \sum_{k=1}^{\infty} k^2 p_k + (\lambda + \mu) \sum_{k=1}^{\infty} k p_k \\ &= 2(\lambda - \mu)\, M_2(N \mid t) + (\lambda + \mu)\, M(N \mid t).\end{aligned}$$

Writing M_2 in place of $M_2(N \mid t)$ for brevity and substituting for $M(N \mid t)$, we now have a differential equation in M_2:

$$\frac{dM_2}{dt} + 2(\mu - \lambda)M_2 = (\lambda + \mu)ie^{(\lambda-\mu)t},$$

which yields

$$e^{2(\mu-\lambda)t}M_2 = i(\lambda+\mu)\int e^{(\lambda-\mu)t}\,e^{2(\mu-\lambda)t}\,dt$$

$$= -\frac{i(\lambda+\mu)}{\lambda-\mu}\,e^{(\mu-\lambda)t} + c_1,$$

where c_1 is the constant of integration. If this constant were evaluated, we should then have an expression for the second moment about the origin of the distribution of N at time t. However, what we set out to determine was not $M_2(N\,|\,t)$ but $\text{var}(N\,|\,t)$, the variance. The variance may be found directly without first evaluating $M_2(N\,|\,t)$, since it differs from it only by a constant, the square of the mean. We may therefore write

$$e^{2(\mu-\lambda)t}\,\text{var}(N\,|\,t) = \frac{-i(\lambda+\mu)}{\lambda-\mu}\,e^{(\mu-\lambda)t} + c_2\,,$$

where c_2 is a different constant. To evaluate it note that at $t=0$, $\text{var}(N\,|\,0)=0$ and thus $c_2 = i(\lambda+\mu)/(\lambda-\mu)$. Therefore

$$\text{var}(N\,|\,t) = e^{2(\lambda-\mu)t}\,\frac{i(\lambda+\mu)}{\lambda-\mu}\,(1-e^{(\mu-\lambda)t})$$

$$= \frac{i(\lambda+\mu)}{\lambda-\mu}\,e^{(\lambda-\mu)t}(e^{(\lambda-\mu)t}-1).$$

Clearly the variance depends not only on the difference between the birth and death rates, that is, on $\lambda-\mu$, the so-called intrinsic rate of natural increase, but also on their absolute magnitudes. This is what we should expect. For a given rate of increase predictions about the future size of the population will be less precise if births and deaths occur in rapid succession than if they are only occasional events.

An alternative way of writing $\text{var}(N\,|\,t)$ is obtained if we put $\lambda-\mu=r$ and $\lambda+\mu=r+2\mu$. Then

$$\text{var}(N\,|\,t) = \frac{i(r+2\mu)}{r}\,e^{rt}(e^{rt}-1).$$

Suppose now that the birth and death rates are equal; that is, $r=0$ and $M(N\,|\,t)=i$ for all t. The variance at time t is now given by $\lim_{r\to 0}\text{var}(N\,|\,t)$. This may be evaluated by using l'Hôpital's rule: differentiating both the numerator and the denominator with respect to r and then putting $r=0$, we see that when $\mu=\lambda$:

$$\text{var}(N\,|\,t) = \lim_{r\to 0}\frac{i[(r+2\mu)(2te^{2rt}-te^{rt})+e^{2rt}-e^{rt}]}{1}$$

$$= 2i\mu t.$$

4. The Chance of Extinction

Consider now the probability that a population governed by a simple birth and death process will die out altogether. The probability that a population will be of size 0 at some time t, given that initially there was only a single individual, is

$$p_0(t \mid i=1) = \frac{\mu e^{(\lambda-\mu)t} - \mu}{\lambda e^{(\lambda-\mu)t} - \mu}.$$

(For a proof of this see Bailey, 1964.)

If a population initially of size i becomes extinct, it follows that each of i separate lines of descent has died out and the probability of this happening is, for any i,

$$p_0(t) = [p_0(t \mid i=1)]^i = \left(\frac{\mu e^{(\lambda-\mu)t} - \mu}{\lambda e^{(\lambda-\mu)t} - \mu}\right)^i. \tag{1.6}$$

To find the probability of ultimate extinction of the population we must now allow t to tend to infinity.

Clearly, if $\lambda < \mu$, the exponential terms vanish as $t \to \infty$ so that $\lim_{t\to\infty} p_0(t) = 1$. Extinction is certain and an indefinitely prolonged persistence of the population is obviously impossible.

If $\lambda > \mu$, then, as $t \to \infty$,

$$p_0(t) \to \left[\frac{\mu e^{(\lambda-\mu)t}}{\lambda e^{(\lambda-\mu)t}}\right]^i = \left(\frac{\mu}{\lambda}\right)^i.$$

Continued existence is not assured, since the probability of extinction remains finite. However, as would be expected, this probability becomes smaller the more the birth rate exceeds the death rate and the larger the size of the initial population.

It remains to consider what will happen if $\lambda = \mu$. To find $\lim_{t\to\infty} p_0(t)$ in this case we may expand the exponentials in (1.6) in the form of series and put $\lambda - \mu = r$. Then

$$p_0(t) = \left[\frac{\mu(rt + r^2t^2/2! + \cdots)}{\lambda(1 + rt + r^2t^2/2! + \cdots) - \mu}\right]^i.$$

As $r \to 0$, terms in r^2 become negligible and

$$p_0(t) \to \left[\frac{\mu rt}{(\lambda-\mu) + \lambda rt}\right]^i \to \left(\frac{\lambda t}{1+\lambda t}\right)^i,$$

since $\lambda - \mu = r$ and $\mu \to \lambda$. Clearly,

$$\lim_{t\to\infty}\left(\frac{\lambda t}{1+\lambda t}\right)^i = 1,$$

from which it follows that ultimate extinction is certain even though the birth and death rates are equal. Although the expected size of the population is a constant, stochastic fluctuations about this expected size will inevitably lead to extinction if a long enough period of time is allowed to pass. Only if $\lambda > \mu$, that is, only if the population has a positive rate of increase, is there a probability (not a certainty) that the population will persist indefinitely.

2

Logistic Population Growth

1. Derivations of the Logistic Equation

In the simple birth and death process it is assumed that the probability that an organism will reproduce or die remains constant and is independent of the size of the population of which it is a member. Obviously this can be true only for a population so small that there is no interference among its members. The growth of any population in a restricted environment must eventually be limited by a shortage of resources. Thus a stage is reached when the demands made by the existing population on these resources preclude further growth and the population is then at its "saturation level," a value determined by the "carrying capacity" of the environment.

Suppose, now, that the growth rate per individual is a function of N, the population size, for all values of N, or that

$$\frac{dN}{dt} = Nf(N).$$

Obviously $df(N)/dN$ must be negative—the larger the population, the greater its inhibitory effect on further growth. The simplest assumption to make is that $f(N)$ is linear or that

$$f(N) = a - bN, \qquad (a, b > 0),$$

whence

$$\frac{dN}{dt} = N(a - bN) \tag{2.1}$$

the well-known Verhulst-Pearl logistic equation.

Because of the basic importance of the logistic equation in population dynamics, it is worth mentioning two other lines of argument that lead to it.

1. Write aN for the potential rate of increase of a population of size N, that is, the rate at which the population would grow if the resources were unlimited and the individuals did not affect one another. Here a is the intrinsic

rate of natural increase denoted by r, or by $\lambda - \mu$, in Chapter 1. Now assume that the actual growth rate is the product of this potential rate times the proportion of the maximum attainable population size that is still unrealized and let this maximum size be a/b. The unrealized proportion when the population is of size N is thus $[(a/b) - N]/(a/b)$ and therefore

$$\frac{dN}{dt} = aN\left[\frac{(a/b) - N}{a/b}\right] = N(a - bN).$$

2. The argument advanced by Lotka (1925) is as follows: We require that the growth rate at any moment should be a function of population size at that moment; that is, we must have $dN/dt = F(N)$. Now suppose that by applying Taylor's theorem $F(N)$ may be expanded as a power series in N so that

$$\frac{dN}{dt} = c_0 + c_1 N + c_2 N^2 + \cdots.$$

Obviously the growth rate is zero at $N = 0$, since at least one individual must be present for the population to grow at all. Thus $c_0 = 0$.

If we were to put $dN/dt = c_1 N$ (the equation for the simple birth and death process), we should have $dN/dt = 0$ only at $N = 0$ and $dN/dt > 0$ for all other population sizes. We now require that $F(N) = 0$ should have two roots, that is, dN/dt should be zero not only for $N = 0$ but also when N reaches its saturation level. The simplest formula for $F(N)$ to satisfy this condition is the one in which the series given above ends with the term in N^2, or

$$F(N) = \frac{dN}{dt} = c_1 N + c_2 N^2.$$

The roots of $F(N)$ are then $N = 0$ and $-c_1/c_2$. Putting $c_1 = a$ and $c_2 = -b$ yields $dN/dt = N(a - bN)$ as before.

The solution of the differential form of the logistic equation is obtained as follows: Rewriting (2.1) in the form of partial fractions, we have

$$\frac{dN}{N} + \frac{b\,dN}{a - bN} = a\,dt.$$

Integrating, we then have

$$\frac{N_t}{a - bN_t} = Ce^{at},$$

where C is a constant of integration or

$$N_t = \frac{Cae^{at}}{1 + Cbe^{at}} = \frac{a/b}{1 + e^{-at}/Cb}.$$

Putting $N_0 = i$ gives $C = i/(a - bi)$ so that

$$N_t = \frac{a}{b}\left[1 + e^{-at}\left(\frac{a/b - i}{i}\right)\right]^{-1}.$$

Alternatively, we may use the form given by Bartlett (1960). Put $Cb = e^{-at_0}$, where t_0 is chosen so that $N_0 = i$. Then

$$N_t = \frac{a/b}{1 + e^{-a(t-t_0)}}. \tag{2.2}$$

The asymptotic value of N_t, namely $\lim_{t\to\infty} N_t = a/b$, is the population's saturation level which it cannot exceed because of environmental limitations.

The form of the logistic curve is shown in Figure 2. There are many examples in the literature of close correspondence between the growth of an actual laboratory population and the theoretical curve. For instance, Gause (1934) describes the growth of a culture of *Paramecium caudatum*, Lotka (1925), an experimental population of *Drosophila* and a colony of bacteria, and Odum (1959), the growth of yeast in a culture. As we shall see, the apparent good fit of the theoretical curve to experimental results may often be spurious. First, however, simple logistic growth is worth investigating in greater detail.

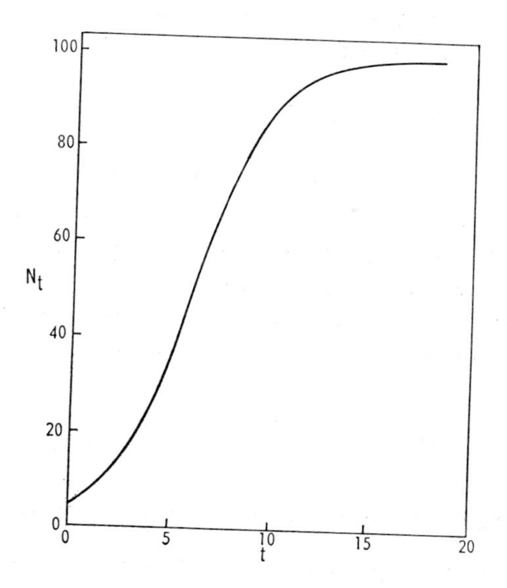

Figure 2. The logistic curve $N_t = (a/b)/[1 + e^{-a(t-t_0)}]$, with $a = 0.5$, $b = 0.005$, $N_0 = 5$, $t_0 = 5.89$.

It is of interest to express (2.2) as a difference equation; that is one that gives population size at time $t+1$ in terms of its size at time t. Rewrite (2.2) as

$$N_t = \frac{a/b}{1 + e^{at_0} e^{-at}} \equiv \frac{K}{1 + J\lambda_1^{-t}}.$$

Here $K = a/b$ is the population's saturation level. The constant e^{at_0} has been replaced by J for brevity and we have put $e^a = \lambda_1$, the finite rate of natural increase; that is, λ_1 is the value of the ratio N_{t+1}/N_t for a population growing without restriction at an intrinsic rate of increase a. Then

$$\begin{aligned} N_{t+1} &= \frac{K}{1 + J\lambda_1^{-(t+1)}} = \frac{K\lambda_1}{\lambda_1 + J\lambda_1^{-t}} \\ &= \frac{\lambda_1 K/(1 + J\lambda_1^{-t})}{1 + (\lambda_1 - 1)/(1 + J\lambda_1^{-t})} \\ &= \frac{\lambda_1 N_t}{1 + N_t(\lambda_1 - 1)/K}. \end{aligned}$$

It should be noticed that the rate of growth given by (2.1) is a net balance of births and deaths. In general, both the birth rate and the death rate will be functions of population size. Denote them by $\lambda(N)$ and $\mu(N)$, respectively. Clearly $\lambda(N)$ must decrease and $\mu(N)$ must increase as N increases; thus we may put $\lambda(N) = a_1 - b_1 N$ and $\mu(N) = a_2 + b_2 N$ with $a_1, a_2 > 0$ and b_1, $b_2 \geq 0$. Equation 2.1 now becomes

$$\begin{aligned} \frac{dN}{dt} &= N[\lambda(N) - \mu(N)] \\ &= N[(a_1 - a_2) - (b_1 + b_2)N]. \end{aligned}$$

Thus the constants a and b in (2.1) each have two components, and it is clear that for given values of a and b there are many possible values for (a_1, b_1) and (a_2, b_2). The form of the logistic curve depends only on the difference between $\lambda(N)$ and $\mu(N)$, not on their separate values. When the population reaches equilibrium at the saturation level, the birth rate and death rate are equal. Then

$$a_1 - b_1 N = a_2 + b_2 N \quad \text{or} \quad N = \frac{a_1 - a_2}{b_1 + b_2} = \frac{a}{b}.$$

2. The Stochastic Form of Logistic Population Growth

So far we have treated logistic growth as though it were deterministic. In fact, of course, births and deaths are chance occurrences. We now consider the stochastic version of the process and how it may be simulated. The events of interest constitute a sequence of births and deaths. Disregarding for the moment the time elapsing between each event and the next, it is clear that at all times there are exactly two possibilities for the next event: it may be a birth, in which case the population size will increase from N to $N+1$; or a death, in which case the population will decrease from N to $N-1$. If all the constants a_1, a_2, b_1, and b_2 are known, we may calculate the probabilities of these events. Thus

$$\Pr(N \to N+1) \propto N\lambda(N) = a_1 N - b_1 N^2$$

and

$$\Pr(N \to N-1) \propto N\mu(N) = a_2 N + b_2 N^2.$$

The events are mutually exclusive and exhaustive, since either a birth or a death must occur next and therefore

$$\Pr(N \to N+1) = \frac{a_1 N - b_1 N^2}{(a_1 + a_2)N - (b_1 - b_2)N^2};$$

$$\Pr(N \to N-1) = \frac{a_2 N + b_2 N^2}{(a_1 + a_2)N - (b_1 - b_2)N^2}.$$

Consider a numerical example. Let the constants be

$$a_1 = 0.7, \qquad a_2 = 0.2,$$

$$b_1 = 0.0045, \qquad b_2 = 0.0005.$$

The deterministic equation is then

$$\frac{dN}{dt} = (a_1 - a_2)N - (b_1 + b_2)N^2 = 0.5N - 0.005N^2,$$

and the equilibrium size of the population is $a/b = 100$. (This is the curve shown in Figure 2.)

We shall now simulate stochastic growth; that is, we shall generate a sequence of births and deaths by assuming these values for the constants and starting with a population of size $N = 70$. First calculate $\Pr(N \to N+1)$ and $\Pr(N \to N-1)$ and denote these probabilities by α and $1-\alpha$, respectively. Now consult a table of random numbers and pick a number in the range $(0, 1]$. If the random number is $\leq \alpha$, let the next event be a birth so that the population increases in size to 71; if the number is $> \alpha$ let the next event be a death so that the population decreases to 69. Once the event has happened and the population size has been adjusted accordingly we may calculate the new probabilities (which depend on the new population size) for the next event and proceed as before.

Table 1 lists a short sequence of events generated in this manner. The value of N in each row, after the first in which it is set equal to 70, is obtained by altering the value in the row above, according as the previous event chanced to be a birth (B) or a death (D).

Thus we have a sequence of events. Notice that as N increases the probability of a death increases and the probability of a birth decreases. The two probabilities become equal when $N = 100$, its equilibrium value.

TABLE 1

N	$\Pr(N \to N+1)$	$\Pr(N \to N-1)$	EVENT (chosen by picking a random number)
70	0.621	0.379	B
71	0.618	0.382	D
70	0.621	0.379	B
71	0.618	0.382	B
72	0.614	0.386	B
73	0.611	0.389	B
74	0.608	0.392	D
73	...		

Next consider *when* the events occur, assuming that simulation begins at time zero. We require the probability density function (pdf) of t, the time to the next event. The probability of an event (of either kind) in an interval of length Δt, when the population is of size N, is $N[\lambda(N) + \mu(N)]\,\Delta t + o(\Delta t)$ or $[(a_1 + a_2)N - (b_1 - b_2)N^2]\,\Delta t + o(\Delta t) \equiv \rho_N\,\Delta t + o(\Delta t)$, say; that is, ρ_N is the rate at which events occur in the population. It is, of course, a function of N, since events will follow one another at shorter intervals in a large population than in a small one. We assume, however, that for each individual the probability that it will reproduce or die is independent of its age. Then, so long as the population remains of size N, the probability of an event in any time interval is independent of the probability in an earlier interval. If we write $p_0(t)$ for the probability that no event will occur in an interval of length t, it then follows that

$$p_0(t + \Delta t) = p_0(t)\,p_0(\Delta t) = p_0(t)(1 - \rho_N\,\Delta t).$$

Then

$$\frac{p_0(t + \Delta t) - p_0(t)}{\Delta t} = -\rho_N\,p_0(t).$$

Letting $\Delta t \to 0$, we have

$$\frac{dp_0(t)}{dt} = -\rho_N \, p_0(t),$$

whence $p_0(t) = Ke^{-\rho_N t}$ with K a constant. Since $p_0(0) = 1, K = 1$ and therefore

$$p_0(t) = e^{-\rho_N t}.$$

This is the probability that by time t no event will have occurred.* The cumulative distribution function of the time elapsing before the next event is thus

$$\begin{aligned} F(t) &= \Pr(\text{time to next event} \leq t) \\ &= 1 - p_0(t) = 1 - e^{-\rho_N t} \end{aligned}$$

and the pdf is $f(t) = F'(t) = \rho_N \, e^{-\rho_N t}$.

For simulation of the process it is necessary to select at random a value of t from the population of values having this pdf. To do this, pick a number, say R, in the range $[0, 1)$ from a random numbers table and set it equal to $F(t)$. Solving for t then yields $t = -(1/\rho_N) \ln (1 - R)$, the random value of t desired. We now simulate in Table 2 the time intervals separating the events shown in Table 1.

TABLE 2

N	ρ_N	RANDOM VALUE OF t	ACCUMULATED TIME	EVENT
70	43.40	0.0005	0.0005	*B*
71	43.74	0.0761	0.0766	*D*
70	43.40	0.0033	0.0799	*B*
71	43.74	0.0025	0.0824	*B*
72	44.06	0.0176	0.1000	*B*
73	44.38	0.0036	0.1036	*B*
74	44.70	0.0400	0.1436	*D*
73	...			

Figure 3 shows the final results of the simulation process by portraying both the events and the times at which they occurred until $t = 0.72$. It is compared with the deterministic curve, which is a short segment of the curve in Figure 2, here enlarged.

* This provides the proof, promised in Chapter 1, that when the death rate is μ the probability of no deaths in an interval of length t is $e^{-\mu t}$.

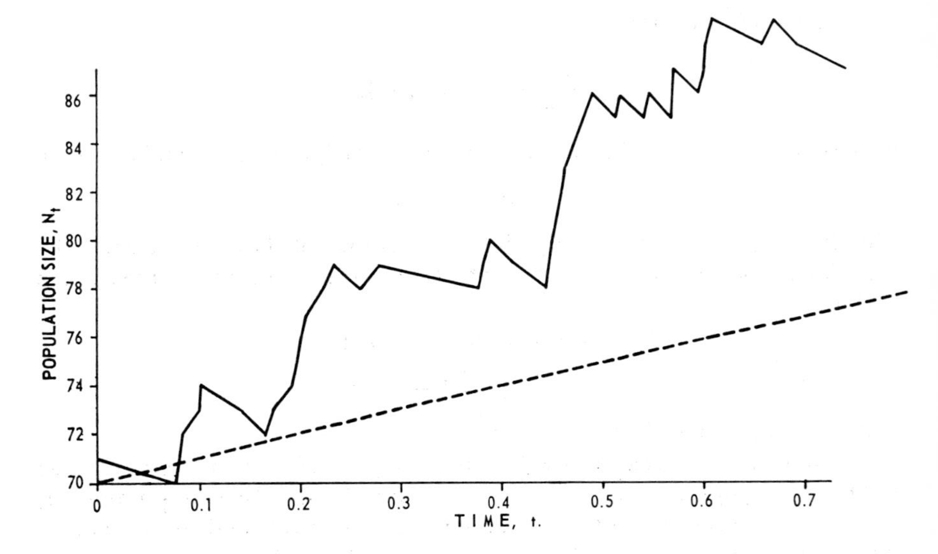

Figure 3. A stochastic realization of logistic population growth. The events (births and deaths) and the times of their occurrence are given, in part, in the text. The broken line shows the corresponding deterministic curve; it appears linear because only a very short segment of it is shown. The constants are $a_1 = 0.7$, $a_2 = 0.2$, $b_1 = 0.0045$, $b_2 = 0.0005$, $N_0 = 70$.

The expected time to the next event in a population of size N is

$$E(t \mid N) = \int_0^\infty tf(t)\,dt = \frac{1}{\rho_N}.$$

In the present example we have

$$E(t \mid N = 70) = \frac{1}{\rho_{70}} = 0.023 \text{ time units}$$

and

$$E(t \mid N = 100) = \frac{1}{\rho_{100}} = 0.020 \text{ time units.}$$

It appears that over this range of values of N the speeding up of events as the population grows is very slight; for this reason the intervals between the events shown do not become noticeably shorter as the population grows.

3. Stochastic Fluctuations at Equilibrium

If we were to allow a stochastic realization of this form of population growth to continue, the population would eventually reach a quasi-stationary state in which its size would hover around the asymptotic equilibrium value indefinitely. The state is *quasi-stationary* and not truly an ultimate stationary state, since extinction always remains possible. Because we have assumed that there is no immigration, we have $\lambda(0) = 0$. This means that when a chance deviation from the equilibrium N is of such magnitude that the size of the population falls to zero, no recovery is possible. However, the probability of this happening may be very small indeed.

If a number of stochastic realizations of the process were allowed to continue for a long time until statistical equilibrium were attained in all of them, we should not expect to find that all the populations were of exactly the same size. Instead, population size at equilibrium will be a random variate and its distribution is what we shall now find.

Clearly, $N\,\mu(N)\,P(N) = (N-1)\lambda(N-1)\,P(N-1)$ when equilibrium has been reached; that is, at equilibrium the probability of a death in a population of size N is the same as the probability of a birth in a population of size $N-1$.

Denoting by N the equilibrium size of the population, we now wish to determine the probabilities $\ldots P(N-2),\ P(N-1),\ P(N),\ P(N+1),\ P(N+2)\ldots$ over the range of population sizes for which the probability is appreciable. It will be found that $P(N-i) \neq P(N+i)$; that is, the distribution is unsymmetrical, although the skewness is slight. The values of $P(N \pm i)$ may be obtained recursively. Begin by choosing a convenient value for $KP(N)$; (the constant K is to be evaluated later). Then $P(N+1)$ may be calculated from

$$K\,P(N+1) = K\,P(N) \cdot \frac{N\,\lambda(N)}{(N+1)\,\mu(N+1)}$$

$$= K\,P(N) \cdot \frac{N(a_1 - b_1 N)}{(N+1)[a_2 + b_2(N+1)]}.$$

In the same way $KP(N+2)$, $KP(N+3)$, $\ldots$, may be found in succession. Similarly,

$$K\,P(N-1) = K\,P(N) \cdot \frac{N(a_2 + b_2 N)}{(N-1)[a_1 - b_1(N-1)]}$$

and $KP(N-2)$, $KP(N-3)$, $\ldots$, may now be calculated. Each series may be prolonged until negligibly small values are reached. The sum of the calculated values is $K\sum_i P(N \pm i) \sim K$, and division of all the calculated terms by K will give the probabilities sought. An example is shown in Figure 4; the parameters are the same as those in the preceding example.

Consider next the properties of the distribution. We must first replace the deterministic equation

$$dN = (aN - bN^2)\,dt$$

with the stochastic equation

$$dN = (aN - bN^2)\,dt + dZ, \tag{2.3}$$

where the first term on the right represents the deterministic part of the change dN, and dZ the chance fluctuation or stochastic part (see Bartlett, 1960).

In what follows we shall need to know the mean and variance of the stochastic displacement dZ. In any small time interval Δt the population size may be displaced from equilibrium either by the occurrence of a single death with probability $N\,\mu(N)\,\Delta t + o(\Delta t)$ or by the occurrence of a single birth with probability $N\lambda(N)\,\Delta t + o(\Delta t)$. These are the probabilities of a displace-

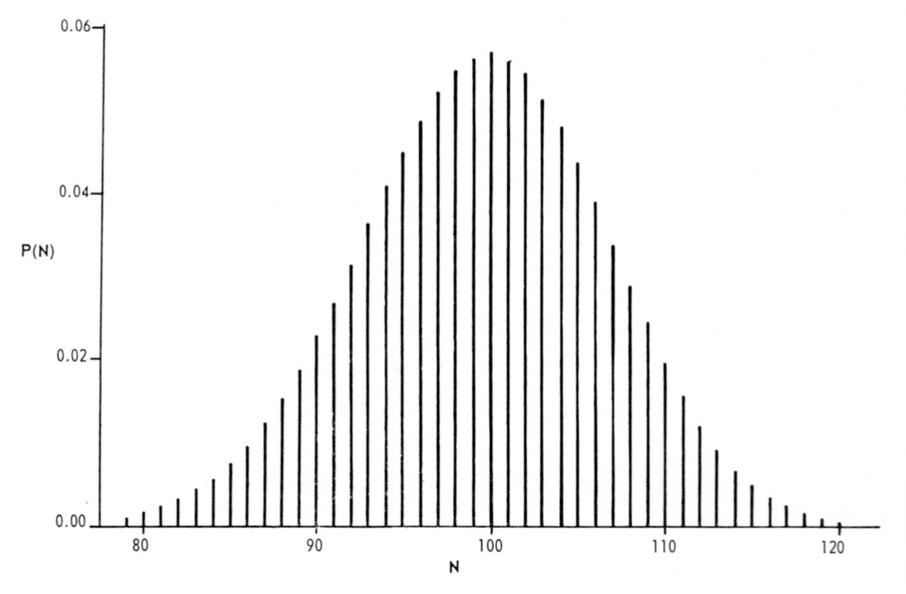

Figure 4. The probability distribution $P(N)$ of the size of a population that has reached stochastic equilibrium. The constants are the same as in Figure 3. The distribution has theoretical moments $\text{var}(N) = \alpha\gamma/2\beta^2 = 50$ and $\hat{m} = 99.50$. The empirical moments are $s^2 = 49.13$ and $\bar{N} = 99.55$.

ment ΔZ of magnitudes -1 and $+1$, respectively. Since, at equilibrium, $\mu(N) = \lambda(N)$, it is seen that $E(\Delta Z) = 0$ and that

$$\text{var}(\Delta Z) = (-1)^2 N\, \mu(N)\, \Delta t + (+1)^2 N\, \lambda(N)\, \Delta t.$$

Thus in the limit $E(dZ) = 0$ and

$$\text{var}(dZ) = N[\lambda(N) + \mu(N)]\, dt = [(a_1 + a_2)N - (b_1 - b_2)N^2]\, dt.$$

We shall now show that the mean of the distribution of N is, because of the asymmetry, somewhat less than the equilibrium value a/b. For suppose the true mean is m and that at time t the population size N deviates from it by an amount X_t; that is, $X_t = N - m$. Then at time $t + dt$ we shall have

$$X_{t+dt} = X_t + (aN - bN^2)\, dt + dZ.$$

Now take expectations: since the unconditional expectations of all the deviations are zero, we have

$$E(X_{t+dt}) = E(X_t) = E(dZ) = 0$$

and consequently $E[(aN - bN^2)\, dt] = 0$ also. Then, since $E(N) = m$ and $E(N^2) = m^2 + \text{var}(N)$,

$$E[(aN - bN^2)\, dt] = \{am - b[m^2 + \text{var}(N)]\}\, dt = 0,$$

whence

$$am = bm^2 + b\, \text{var}(N) \quad \text{or} \quad m = \frac{a}{b} - \frac{\text{var}(N)}{m}.$$

As an estimator of m we may take $\hat{m} = a/b - (b/a)\, \text{var}(N)$, since $m \sim a/b$.

Next, consider $\text{var}(N)$. Assume that N departs only slightly from its equilibrium value of a/b so that we may write $N = (a/b)(1 + u)$ with u small. Then $dN = (a/b)\, du$ and the stochastic equation (2.3) becomes

$$\frac{a}{b} \cdot du = \left[\frac{a^2}{b}(1 + u) - b\frac{a^2}{b^2}(1 + u)^2\right] dt + dZ$$

or

$$du = -au(1 + u)\, dt + \frac{b}{a}\, dZ \sim -au\, dt + \frac{b}{a}\, dZ$$

since $1 + u \sim 1$. Then

$$(u + du)^2 = u^2(1 - a\, dt)^2 + 2u\frac{b}{a}(1 - a\, dt)\, dZ + \left(\frac{b}{a}\, dZ\right)^2.$$

Take expectations, bearing in mind (a) that $E(dZ) = 0$ and $E[(dZ)^2] = \text{var}(dZ)$ and (b) that since we are dealing with a state of stochastic equilibrium the quantity u has the same properties at times t and $t + dt$ or equivalently that $E(u + du)^2 = E(u^2) = \text{var}(u)$.

Therefore, neglecting terms in $(dt)^2$,

$$\text{var}(u) = (1 - 2a\, dt)\, \text{var}\,(u) + \frac{b^2}{a^2}\, \text{var}(dZ)$$

$$= (1 - 2a\, dt)\, \text{var}\,(u) + \gamma\, dt,$$

where

$$\gamma = \frac{b^2}{a^2}[(a_1 + a_2)N - (b_1 - b_2)N^2] \sim \frac{b}{a}\,(a_1 + a_2) - (b_1 - b_2),$$

since $N \sim a/b$. Solving for var(u), we find that $\text{var}(u) = \gamma/2a$ and consequently that

$$\text{var}(N) \sim \text{var}\left(\frac{a}{b}\, u\right) = \frac{a^2}{b^2} \cdot \frac{\gamma}{2a} = \frac{a\gamma}{2b^2}.$$

For the numerical example already described the moments are

$$\text{var}(N) = \frac{a\gamma}{2b^2} = 50.$$

and

$$\hat{m} = \frac{a}{b} - \frac{b}{a}\, \text{var}(N) = 99.50.$$

Evaluating the moments of the calculated distribution shown in Figure 4 gives

$$\bar{N} = 99.55 \quad \text{and} \quad \text{var}(N) = 49.13.$$

The agreement is very good.

4. A Modification of the Logistic Equation

Although the logistic curve is often fitted to observed growth curves of natural and laboratory populations and sometimes appears to fit well, it is worth keeping in mind the highly simplified assumptions underlying it. Six assumptions that must frequently be false are the following:

1. Abiotic environmental factors are sufficiently constant not to affect the birth and death rates.

2. Crowding affects all population members equally. This is unlikely to be true if the individuals occur in clumps instead of being evenly distributed throughout the available space.

3. Birth and death rates respond instantly, without lag, to density changes.

4. Population growth rate is density-dependent even at the lowest densities. It may be more reasonable to suppose that there is some threshold density below which individuals do not interfere with one another.

5. The population has, and maintains, a stable age distribution.

6. The females in a sexually reproducing population always find mates, even when the density is low.

If assumptions (4) and (6) were incorrect, the resultant deviations would tend to cancel each other out, but it is hardly likely that the effects would ever balance precisely.

It seems probable that often the good fit of the logistic curve to observed growth curves is apparent rather than real. Observed growth curves are usually sigmoid. If the only observations made on a population are determinations of its size at a succession of times and if these are the data used to estimate the parameters a and b of the logistic equation, we can scarcely avoid getting at least a fairly good fit. Other data, besides those obtained by merely recording the growth of an undisturbed population, are needed before a conclusion can be drawn that the growth of a particular population is, or is not, logistic.

As an example, consider the investigations of Smith (1963) on laboratory populations of *Daphnia magna*. An "expanding culture" technique was used in growing the populations by which a population could be maintained indefinitely at a given level of crowding. The growth rates at a succession of chosen density levels, imposed by the experimenter, were then determined.

Now consider the logistic equation in the form

$$\frac{1}{N}\frac{dN}{dt} = a\left(\frac{K - N}{K}\right), \tag{2.4}$$

where we have written $K = a/b$ for the saturation level. Clearly, if this equation describes population growth, the growth rate per individual $(1/N)\, dN/dt$ decreases linearly with increasing N. In his experiments, however, Smith found that the relation between these quantities was not a straight line but a concave curve.

He then argued as follows. In its usual form (2.4) the logistic equation contains the assumption that growth rate is proportional to $(K - N)/K$, the proportion of maximum attainable population size still unrealized (see page 20). However, it might be truer to say that growth rate depends on the proportion of some limiting factor not yet utilized. For a food-limited population the term $(K-N)/K$ should then be replaced with a term representing the proportion of "the rate of food supply not momentarily being used by the population" (Smith, 1963). Notice that the *rate* of food supply is considered, not the amount of food. We may then write

$$\frac{1}{N}\frac{dN}{dt}=a\left(\frac{T-F}{T}\right), \tag{2.5}$$

where F is the rate at which a population of size N uses food and T is the corresponding rate when the population reaches saturation level. The ratio F/T is not the same as N/K, since a growing population will use food faster than a saturated population.* When the population is growing, food is used for both maintenance and growth, whereas once the saturation level has been reached and no further growth takes place it is used for maintenance only. Thus we must have $F/T > N/K$. Now, F must depend on N (the size of the population being maintained) and dN/dt (the rate at which the population is growing), and therefore the simplest assumption to make concerning F is that

$$F=c_1N+c_2\frac{dN}{dt}, \qquad (c_1, c_2>0).$$

When saturation is reached, $dN/dt=0$, and by definition $N=K$ and $F=T$, whence we obtain $T=c_1K$.

Equation 2.5 may now be rewritten

$$\frac{1}{N}\cdot\frac{dN}{dt}=a\left(\frac{c_1K-c_1N-c_2\,dN/dt}{c_1K}\right)$$

or, putting $c_1/c_2=c$,

$$\frac{1}{N}\cdot\frac{dN}{dt}=a\left[\frac{K-N}{K+(a/c)N}\right] \tag{2.6}$$

and this may be contrasted with (2.4). It will be seen that if (2.6) holds the growth rate per individual will decrease rapidly with increasing N while N is small and less rapidly as N becomes larger. This is what Smith found to be true in populations of *Daphnia magna*.

* Smith uses M, the mass, rather than N, the number of individuals, as a measure of population size. Here N is used simply to avoid the additional complication of a change of units.

3

Population Growth with Age-Dependent Birth and Death Rates I: The Discrete Time Model

1. Introduction

So far we have assumed either that birth and death rates were independent of age or that population changes occurred in such a way that the age distribution remained unaltered. In arriving at the logistic equation, we did, however, allow for density dependence or a decrease in birth rate and an increase in death rate as the population became larger.

We shall now reverse these assumptions: that is, we shall assume that an individual's chances of reproducing and dying are a function of its age but that these chances are unaffected by the size of the population in which it finds itself, perhaps because population numbers never rise so high that density dependence begins to exert any influence. We shall be concerned to discover not merely the size of a population after a lapse of time, given the initial size, but also the age distribution within the population after a lapse of time, given the initial age distribution. Natural realizations of the process to be described would occur, for instance, if there were sudden colonization of a new area by a small immigrant population of arbitrary age distribution. The model considered here is deterministic.

It simplifies discussion to consider only the females in a bisexual population. The same arguments would also apply if one counted members of both sexes, provided the ratio of males to females remained constant and that at all ages the death rates were the same for both sexes. In this chapter, following Lewis (1942) and Leslie (1945, 1948), we consider a discrete time model with a discrete age scale and assume that *within* an age interval birth and death rates remain constant; they differ from one interval to the next. The predictions are therefore not exact but if the time intervals are short the approximation is good. The model treating time as continuous, and with a continuous age scale, is discussed in Chapter 4.

Assume that at time $t = 0$ a population can be represented by the column vector

$$\mathbf{n}_0 = \begin{pmatrix} n_{00} \\ n_{10} \\ n_{20} \\ \vdots \\ n_{m0} \end{pmatrix}.$$

There are $m + 1$ elements in the vector, that is, $m + 1$ different age groups. The first subscript denotes age and the second, time. Thus at time t there are n_{xt} individuals (females) in the age range x to $x + 1$ time units (equivalent to age x last "birthday"). Such a female is described as of age x. It is assumed that none can live to be older than m. The time intervals have the same duration as the age intervals.

Now denote by F_x the number of daughters born in one unit of time to a female of age x that will survive into the next unit of time. Until the first unit of time has passed their ages will be 0. Also, write P_x for the probability that a female aged x at time t will survive to time $t + 1$; her age will then be $x + 1$. So $P_x n_{xt} = n_{x+1,\, t+1}$.

Beginning, then, with a population described by the vector given above, we now ask what the population will consist of one time unit later, at $t = 1$. Clearly, in matrix notation, the change in a unit of time can be represented as follows:

$$\begin{pmatrix} F_0 & F_1 & \cdots & F_{m-1} & F_m \\ P_0 & 0 & \cdots & 0 & 0 \\ 0 & P_1 & \cdots & 0 & 0 \\ \cdots & \cdots & \cdots & \cdots & \cdots \\ 0 & 0 & \cdots & P_{m-1} & 0 \end{pmatrix} \begin{pmatrix} n_{00} \\ n_{10} \\ n_{20} \\ \vdots \\ n_{m0} \end{pmatrix} = \begin{pmatrix} F_0 n_{00} + \cdots + F_m n_{m0} \\ P_0 n_{00} \\ P_1 n_{10} \\ \vdots \\ P_{m-1} n_{m-1,0} \end{pmatrix} = \begin{pmatrix} n_{01} \\ n_{11} \\ n_{21} \\ \vdots \\ n_{m1} \end{pmatrix}$$

or

$$\mathbf{M}\mathbf{n}_0 = \mathbf{n}_1;$$

$\mathbf{M}$ may be called a "projection matrix." Similarly

$$\mathbf{n}_2 = \mathbf{M}\mathbf{n}_1 = \mathbf{M}^2\mathbf{n}_0; \ldots; \mathbf{n}_t = \mathbf{M}^t\mathbf{n}_0,$$

where $\mathbf{n}_t$ denotes the population structure after t time units have elapsed. Note that $\mathbf{M}$ is an $(m + 1) \times (m + 1)$ matrix. Unless the females are reproductive until the end of their lifespan, some of the elements in the first row of $\mathbf{M}$ (at the right-hand end when the females are in their postreproductive stage) may be zero. Then $|M| = 0$ and the matrix is singular. Thus suppose the females are sterile in the last $m - k$ age units so that **M** may be written

$$\left(\begin{array}{ccccc|cccc}
F_0 & F_1 & \cdots & F_{k-1} & F_k & 0 & \cdots & 0 & 0 \\
P_0 & 0 & \cdots & 0 & 0 & 0 & \cdots & 0 & 0 \\
0 & P_1 & \cdots & 0 & 0 & 0 & \cdots & 0 & 0 \\
\cdots & \cdots & \cdots & \cdots & \cdots & \cdots & \cdots & \cdots & \cdots \\
0 & 0 & \cdots & P_{k-1} & 0 & 0 & \cdots & 0 & 0 \\
\hline
\cdots & \cdots & \cdots & \cdots & \cdots & \cdots & \cdots & \cdots & \cdots \\
0 & 0 & \cdots & 0 & 0 & 0 & \cdots & P_{m-1} & 0
\end{array}\right)$$

$$\equiv \left(\begin{array}{c|c}
\mathbf{A} & \mathbf{0} \\
(k+1) \times (k+1) & (k+1) \times (m-k) \\
\hline
\mathbf{B} & \mathbf{C} \\
(m-k) \times (k+1) & (m-k) \times (m-k)
\end{array}\right).$$

Then

$$\mathbf{M}^t = \begin{pmatrix} \mathbf{A}^t & \mathbf{0} \\ f(\mathbf{ABC}) & \mathbf{C}^t \end{pmatrix}.$$

Now **C** is strictly triangular; its only nonzero elements, $P_{k+1}, \ldots, P_{m-1}$, are on the subdiagonal. Therefore, when $t \geq m - k$, $\mathbf{C}^t = \mathbf{0}$ and $\mathbf{M}^t$ has only zeroes in its last $m - k$ columns. This is a mathematical statement of the fact that those females that were already of postreproductive age at $t = 0$ are unrepresented, either in person or by descendents, once they are dead.

Now consider **A** only. Repeated premultiplication by **A** of the $(k+1)$-element column vector $(n_{00}\ n_{10}\ \cdots\ n_{k0})'$ permits prediction of the age distribution of that part of the population of reproductive age. In what follows we shall disregard females of postreproductive age.

A is nonsingular (even if there are *pre*reproductive sterile stages) and has determinant

$$|A| = (-1)^{k+2}(F_k P_0 P_1 \cdots P_{k-1}).$$

Therefore it has an inverse $\mathbf{A}^{-1}$, and if we are given a vector $\mathbf{n}_t$ to describe the age distribution at time t we may find the age distribution at successively ealier times $t-1$, $t-2, \ldots$, from $\mathbf{n}_{t-1} = \mathbf{A}^{-1}\mathbf{n}_t$; $\mathbf{n}_{t-2} = A^{-2}\mathbf{n}_t$; $\cdots$. Now,

$$\mathbf{A}^{-1} = \begin{pmatrix}
0 & P_0^{-1} & 0 & 0 & \cdots & 0 \\
0 & 0 & P_1^{-1} & 0 & \cdots & 0 \\
0 & 0 & 0 & P_2^{-1} & \cdots & 0 \\
\cdots & \cdots & \cdots & \cdots & \cdots & \cdots \\
\dfrac{1}{F_k} & \dfrac{-F_0}{P_0 F_k} & \dfrac{-F_1}{P_1 F_k} & \dfrac{-F_2}{P_2 F_k} & \cdots & \dfrac{-F_{k-1}}{P_{k-1} F_k}
\end{pmatrix}.$$

It is clear that repeated premultiplication by $\mathbf{A}^{-1}$, which has k negative elements in the last row, may eventually give a negative value for the last element $n_{kt'}$, say, at some time t' preceding t. This is obviously impossible: it would imply a negative number of females in the oldest reproductive age group at some earlier time.

Therefore, although the forward series can be generated indefinitely into the future, the backward operation can be performed only until the stage preceding that for which n_{kt} becomes negative. The explanation is as follows: if all our assumptions hold and if, as we have also assumed, the process is deterministic, the fate of *any* arbitrary initial age distribution can be predicted as far forward in time as we like; the starting distribution may be whatever we care to choose. But this same arbitrary distribution may be an impossible outcome for a long-continued process; that is, there may be *no* initial age distribution such that if the forward operation is repeatedly applied to it our arbitrary distribution will be at some time result.

Returning to a consideration of the forward projection matrix $\mathbf{A}$, it is worth noting that by its use we can predict the contribution of any particular age group, say those of age j at $t = 0$, to the population at some future time t.

Let $\mathbf{A}^t = \{\phi_{st}\}$, $\quad s, t = 0, 1, \ldots, k$. Then $\mathbf{A}^t\mathbf{n}_0$ becomes

$$\begin{pmatrix} \phi_{00} & \phi_{01} & \cdots & \phi_{0k} \\ \phi_{10} & \phi_{11} & \cdots & \phi_{1k} \\ \cdots & \cdots & \cdots & \cdots \\ \phi_{k0} & \phi_{k1} & \cdots & \phi_{kk} \end{pmatrix} \begin{pmatrix} n_{00} \\ n_{10} \\ \vdots \\ n_{k0} \end{pmatrix} = \begin{pmatrix} \cdots + \phi_{0j}\, n_{j0} + \cdots \\ \cdots + \phi_{1j}\, n_{j0} + \cdots \\ \vdots \\ \cdots + \phi_{kj}\, n_{j0} + \cdots \end{pmatrix} = \begin{pmatrix} n_{0t} \\ n_{1t} \\ \vdots \\ n_{kt} \end{pmatrix},$$

where we have shown only those contributions to the elements in $\mathbf{n}_t$ that come from the element n_{j0} of $\mathbf{n}_0$. At time t, provided $t > k$, every element in $\mathbf{n}_t$ has some contribution from the n_{j0} individuals that were of age j at the start. At time t the total number of individuals that are survivors or descendents of this original age group is given by

$$n_{j0} \sum_{i=0}^{k} \phi_{ij}\,;$$

that is, the product of n_{j0} (the number in the ancestral age group we are concerned with) and the sum of the elements in the $(j + 1)$th column of $\mathbf{A}^t$.

2. The Stable Age Distribution

Excluding the postreproductive age groups, that is, considering only the matrix $\mathbf{A}$, it is clear that there may be an age distribution vector $\mathbf{n}_s$, say, such that

$$\mathbf{n}_{s+1} = \mathbf{A}\mathbf{n}_s = \lambda\mathbf{n}_s\,,$$

where λ is some scalar constant. The elements of $\mathbf{n}_{s+1}$ are proportional to those in $\mathbf{n}_s$, though at the same time the size of the whole population at time $s+1$ is λ times its size at time s. Such an age distribution is *stable* and we call $\mathbf{n}_s$ the vector of the stable age distribution or the stable vector for short. The whole population will also be *stationary* in size if $\lambda = 1$. But we need not have $\lambda = 1$; the population as a whole may grow or dwindle while still maintaining its stable age distribution.

Given $\mathbf{A}$, we now wish to find the stable vector $\mathbf{n}_s$; that is, we wish to solve

$$\mathbf{A}\mathbf{n}_s = \lambda \mathbf{n}_s . \tag{3.1}$$

The result will be, of course, the proportions of the different age classes in a stable population of any size. To solve (3.1) explicity it is convenient to transform the matrices from their original coordinate system to a new frame of coordinates. The equation $\mathbf{A}\mathbf{n}_s = \lambda\mathbf{n}_s$ may be difficult to solve, but if we can write an equivalent equation in transformed variables, say $\boldsymbol{\nu}_{s+1} = \mathbf{B}\boldsymbol{\nu}_s = \lambda\boldsymbol{\nu}_s$, and if $\mathbf{B}$ is of simple form, the equation may be made easily solvable. Here $\mathbf{B}$ operating on $\boldsymbol{\nu}_s$ in a new coordinate frame is equivalent to $\mathbf{A}$ acting on $\mathbf{n}_s$ in the old frame. The desired transformation may be found as follows:

First, put $\boldsymbol{\nu}_s = \mathbf{H}\mathbf{n}_s$, where

$$\mathbf{H} = \begin{pmatrix} \prod_0^{k-1} P_i & 0 & \cdots & 0 & 0 \\ 0 & \prod_1^{k-1} P_i & \cdots & 0 & 0 \\ \cdots & \cdots & \cdots & \cdots & \cdots \\ 0 & 0 & \cdots & P_{k-1} & 0 \\ 0 & 0 & \cdots & 0 & 1 \end{pmatrix}.$$

Then $\boldsymbol{\nu}_{s+1} = \mathbf{H}\mathbf{n}_{s+1} = \mathbf{H}\mathbf{A}\mathbf{n}_s = (\mathbf{H}\mathbf{A}\mathbf{H}^{-1})\boldsymbol{\nu}_s \equiv \mathbf{B}\boldsymbol{\nu}_s$, say, where $\mathbf{B} = \mathbf{H}\mathbf{A}\mathbf{H}^{-1}$. Thus

$$\mathbf{B} = \begin{pmatrix} F_0 & P_0F_1 & P_0P_1F_2 & \cdots & (P_0P_1\cdots P_{k-1})F_k \\ 1 & 0 & 0 & \cdots & 0 \\ 0 & 1 & 0 & \cdots & 0 \\ \cdots & \cdots & \cdots & \cdots & \cdots \\ 0 & 0 & 0 & \cdots\ 1 & 0 \end{pmatrix}.$$

Next, to find λ we note that since, by definition, $\mathbf{B}\boldsymbol{\nu}_s = \boldsymbol{\lambda}\boldsymbol{\nu}_s$ or $(\mathbf{B} - \lambda\mathbf{I})\boldsymbol{\nu}_s = \mathbf{0}$, λ may be found by solving the equation $|\mathbf{B} - \lambda\mathbf{I}| = 0$.

Expanding the determinant gives

$$\lambda^{k+1} - F_0\lambda^k - P_0F_1\lambda^{k-1} - P_0P_1F_2\lambda^{k-2} - \cdots - (P_0P_1\cdots P_{k-1})F_k = 0. \tag{3.2}$$

Notice that the coefficients of λ^k, λ^{k-1}, ..., are the elements in the first row of **B** written with minus signs. Since the left-hand side of this equation has only one change of sign, only one of the roots will be real and positive.* Call this dominant root λ_1.

We now solve $\mathbf{B}\boldsymbol{\nu}_s = \lambda_1 \boldsymbol{\nu}_s$ to obtain the values ν_{xs} ($x = 0, 1, \ldots, k$), which are the elements of the stable vector in the new coordinate frame. In fact, since **B** has 1's on the subdiagonal and 0's elsewhere, except in its first row,

$$\begin{pmatrix} b_{00} & b_{01} & \cdots & b_{0,k-1} & b_{0k} \\ 1 & 0 & \cdots & 0 & 0 \\ 0 & 1 & \cdots & 0 & 0 \\ \cdots & \cdots & \cdots & \cdots & \cdots \\ 0 & 0 & \cdots & 1 & 0 \end{pmatrix} \begin{pmatrix} \nu_{0s} \\ \nu_{1s} \\ \nu_{2s} \\ \vdots \\ \nu_{ks} \end{pmatrix} = \lambda_1 \begin{pmatrix} \nu_{0s} \\ \nu_{1s} \\ \nu_{2s} \\ \vdots \\ \nu_{ks} \end{pmatrix},$$

(where we have put b_{0i} ($i = 0, \ldots, k$) in the first row of **B**) and therefore

$$\nu_{0s} = \lambda_1 \nu_{1s},\ \nu_{1s} = \lambda_1 \nu_{2s},\ \ldots,\ \nu_{k-1,s} = \lambda_1 \nu_{ks}.$$

Thus the stable vector in the new coordinate frame is

$$\boldsymbol{\nu}_s \propto \begin{pmatrix} \lambda_1{}^k \\ \lambda_1^{k-1} \\ \vdots \\ \lambda_1 \\ 1 \end{pmatrix}.$$

To obtain the stable vector $\mathbf{n}_s$ in the original coordinate frame it is now necessary only to note that

$$\mathbf{n}_s = \mathbf{H}^{-1}\boldsymbol{\nu}_s = \begin{pmatrix} \lambda_1{}^k/(P_0 P_1 \cdots P_{k-1}) \\ \lambda_1^{k-1}/(P_1 P_2 \cdots P_{k-1}) \\ \lambda_1^{k-2}/(P_2 P_3 \cdots P_{k-1}) \\ \vdots \\ \lambda_1/P_{k-1} \\ 1 \end{pmatrix}. \tag{3.3}$$

This is the stable age distribution. Summarizing, we have done the following: using the elements of **A**, we have set up (3.2) in order to obtain λ_1, the rate of increase per unit of time when stability has been achieved. Using λ_1, we have then obtained the stable age distribution from (3.3) in which the elements of $\mathbf{n}_s$ are expressed in terms of λ_1 and the elements of **A**.

* This follows from Descartes' rule of signs. For a proof see, for instance, C. C. MacDuffee (1954), *Theory of Equations*, Wiley, New York.

A numerical example (artificial) will clarify this. (The numbers are unrealistic and have been chosen to keep the arithmetic simple.)

$$\mathbf{A} = \begin{pmatrix} 1 & 3 & 4 & 12 \\ \frac{1}{2} & 0 & 0 & 0 \\ 0 & \frac{1}{4} & 0 & 0 \\ 0 & 0 & \frac{2}{3} & 0 \end{pmatrix}.$$

Then

$$\mathbf{H} = \begin{pmatrix} \frac{1}{12} & 0 & 0 & 0 \\ 0 & \frac{1}{6} & 0 & 0 \\ 0 & 0 & \frac{2}{3} & 0 \\ 0 & 0 & 0 & 1 \end{pmatrix} \quad \text{and} \quad \mathbf{B} = \begin{pmatrix} 1 & \frac{3}{2} & \frac{1}{2} & 1 \\ 1 & 0 & 0 & 0 \\ 0 & 1 & 0 & 0 \\ 0 & 0 & 1 & 0 \end{pmatrix};$$

hence

$$|\mathbf{B} - \lambda\mathbf{I}| = \begin{vmatrix} 1-\lambda & \frac{3}{2} & \frac{1}{2} & 1 \\ 1 & -\lambda & 0 & 0 \\ 0 & 1 & -\lambda & 0 \\ 0 & 0 & 1 & -\lambda \end{vmatrix} = 0$$

or $\lambda^4 - \lambda^3 - \frac{3}{2}\lambda^2 - \frac{1}{2}\lambda - 1 = 0$; $\lambda_1 = 2$ is the only positive real root.

Now $\mathbf{B}\boldsymbol{\nu}_s = \lambda_1 \boldsymbol{\nu}_s$ and $\lambda_1 = 2$. Therefore

$$\begin{pmatrix} 1 & \frac{3}{2} & \frac{1}{2} & 1 \\ 1 & 0 & 0 & 0 \\ 0 & 1 & 0 & 0 \\ 0 & 0 & 1 & 0 \end{pmatrix} \begin{pmatrix} \nu_{0s} \\ \nu_{1s} \\ \nu_{2s} \\ \nu_{3s} \end{pmatrix} = 2 \begin{pmatrix} \nu_{0s} \\ \nu_{1s} \\ \nu_{2s} \\ \nu_{3s} \end{pmatrix},$$

whence

$$\nu_{0s} = 2\nu_{1s}; \quad \nu_{1s} = 2\nu_{2s}; \quad \nu_{2s} = 2\nu_{3s}.$$

So, putting $\nu_{3s} = 1$, we have $\nu_{2s} = 2$; $\nu_{1s} = 4$; $\nu_{0s} = 8$. Thus, in transformed coordinates, the stable vector is

$$\boldsymbol{\nu}_s \propto \begin{pmatrix} 8 \\ 4 \\ 2 \\ 1 \end{pmatrix}.$$

Converting back to the original coordinate frame gives

$$\mathbf{n}_s = \mathbf{H}^{-1}\boldsymbol{\nu}_s$$

or

$$\mathbf{n}_s = \begin{pmatrix} n_{0s} \\ n_{1s} \\ n_{2s} \\ n_{3s} \end{pmatrix} \propto \begin{pmatrix} 12 & 0 & 0 & 0 \\ 0 & 6 & 0 & 0 \\ 0 & 0 & \frac{3}{2} & 0 \\ 0 & 0 & 0 & 1 \end{pmatrix} \begin{pmatrix} 8 \\ 4 \\ 2 \\ 1 \end{pmatrix} = \begin{pmatrix} 96 \\ 24 \\ 3 \\ 1 \end{pmatrix}.$$

It will be seen that, as required, $\mathbf{A}\mathbf{n}_s = \lambda_1 \mathbf{n}_s$ or $2\mathbf{n}_s$. We also see that substituting λ_1 and the P_i values given in $\mathbf{A}$ into (3.3) gives the same result.

It has also been shown by Leslie (1945) that whatever the initial age distribution it will tend to the stable form after sufficiently many generations if $\lambda_1 \leq 1$ (unless, of course, the population has died out before stability is reached). The same is true if $\lambda_1 > 1$, provided none of the other roots of the equation $|\mathbf{B} - \lambda\mathbf{I}| = 0$ has a modulus greater than 1. If the latter condition does not hold, however, that is, if, in addition to λ_1, there is another root (or roots) of modulus >1, the population may, depending on its initial age distribution, continue to deviate from the stable form indefinitely. Details will be found in Leslie's papers (1945, 1948).

Some combinations of matrix and initial vector produce regular cyclic repetitions of this vector; for example, suppose $\mathbf{A}$ were of the form

$$\mathbf{A} = \begin{pmatrix} 0 & 0 & \cdots & 0 & F \\ P_0 & 0 & \cdots & 0 & 0 \\ 0 & P_1 & \cdots & 0 & 0 \\ \cdots & \cdots & \cdots & \cdots & \cdots \\ 0 & 0 & \cdots & P_{k-1} & 0 \end{pmatrix},$$

as might be true of population growth in an insect species that is fertile only in an adult stage of very short duration. It is easy to see that whatever the initial age distribution it will repeat itself cyclically with cycles of length $k+1$. At the end of a cycle the population's size will be $(FP_0 P_1 \cdots P_{k-1})$ times what it was at the beginning of the cycle.

In all of the foregoing arguments it is assumed that the age intervals are of equal length and that, in judging how well an actual population conforms to the model, one can tell at a glance to which age group any individual belongs. In ecological work it is usually impossible to judge an individual's age, but many animals (e.g., insects) go through a sequence of easily recognizable stages that are of unequal duration. Lefkovitch (1965) has generalized Leslie's model by considering unequal stage groupings instead of equal age groupings; further, no assumptions are made about possible variation in stage duration in different individuals. Lefkovitch demonstrated the usefulness of his model in studies of the growth of experimental populations of the cigarette beetle, *Lasioderma serricorne* (Fabricius).

Pollard (1966) has considered the stochastic version of the deterministic theory developed by Leslie. He has also shown that predictions based on the model are not markedly affected by changes in the length of the discrete time interval. Thus he compared two estimates of the rate of increase per annum of Australian human females and showed that the estimates were very close, regardless of whether the age classes were one year or five years long. In theory lumping of the age classes to produce a coarser age classification should affect the result, for if the process is Markovian at a chosen time interval it will, in general, be non-Markovian when a different time interval is used. However, the error introduced by altering the time interval appears to be small.

3. The Effect of Density Dependence on Leslie's Model

Leslie (1959) proposed a modified form of projection matrix to allow for the effect on population growth of the presence of other population members. Consider again the nonsingular matrix **A** already described (see page 35). As before, let its dominant latent root be λ_1. Thus, if $\mathbf{n}_s$ is a vector proportional to the stable age distribution, $\mathbf{A}\mathbf{n}_s = \lambda_1 \mathbf{n}_s$. Notice that the $(i+1)$th column of **A** gives the fertility and survival rates of the ith age group. Now let each element in this column be divided by q_{it} for $i = 0, 1, \ldots, k$, where

$$q_{it} = 1 + aN_{t-i-1} + bN_t .$$

Here $a, b > 0$ are constants that are the same for all age classes; consequently, $q_{it} > 1$. The projection matrix now becomes a function of time and is

$$\mathbf{A}_t = \mathbf{A}\mathbf{Q}_t^{-1},$$

where

$$\mathbf{Q}_t = \begin{pmatrix} q_{0t} & 0 & \cdots & 0 \\ 0 & q_{1t} & \cdots & 0 \\ \cdots & \cdots & \cdots & \cdots \\ 0 & 0 & \cdots & q_{kt} \end{pmatrix}.$$

Then $\mathbf{A}_t \mathbf{n}_t = \mathbf{A}\mathbf{Q}_t^{-1}\mathbf{n}_t = \mathbf{n}_{t+1}$.

It will be seen that the rates of production of surviving young (the elements F_i in **A**) and also the survivorship of each age group (the elements P_i in **A**) have been decreased in defining $\mathbf{A}_t$. Each element in **A** has been divided by an amount that depends on (a) the size of the current population at time t and (b) the size of the population at time $t - i - 1$, which is the beginning of the time interval in which individuals currently of age i were born. We are assuming, therefore, that the birth and death rates of all population members are influenced both by their present degree of crowding and by the crowding they experienced at the beginnings of their lifetimes which might well have a lasting effect on their future chances of reproducing and surviving.

The population will become stationary, that is, its age distribution will be stable and its total size will remain constant when, at some time τ, $q_{it} = \lambda_1$ for all i. Denoting the equilibrium size of the population by K, we then have

$$N_\tau = N_{\tau+1} = \cdots = N_{\tau+k+1} = \cdots = K.$$

Also, since $q_{i,\tau+k+1} = \lambda_1 = 1 + aN_{\tau+k-i} + bN_{\tau+k+1}$ $(i = 0, 1, \ldots, k)$,
$$= 1 + (a + b)K,$$

it follows that

$$K = \frac{\lambda_1 - 1}{a + b}.$$

At this stage

$$\mathbf{A}_\tau = \mathbf{A}\mathbf{Q}_\tau^{-1} = \lambda_1^{-1}\mathbf{A}.$$

Thus, when the population has become stationary, its projection matrix is

$$\mathbf{A} = \begin{pmatrix} \frac{F_0}{\lambda_1} & \frac{F_1}{\lambda_1} & \cdots & \frac{F_{k-1}}{\lambda_1} & \frac{F_k}{\lambda_1} \\ \frac{P_0}{\lambda_1} & 0 & \cdots & 0 & 0 \\ 0 & \frac{P_1}{\lambda_1} & \cdots & 0 & 0 \\ \cdots & \cdots & \cdots & \cdots & \cdots \\ 0 & 0 & & \frac{P_{k-1}}{\lambda_1} & 0 \end{pmatrix}.$$

The stable age distribution in this density-dependent population is seen to be the same as that in a population having the original **A** as its projection matrix, but the population's total size remains stationary.

Leslie (1959) gives a numerical example of this form of population growth and the result is shown in Figure 5. The population exhibits damped oscillations as it gradually approaches the stationary state.

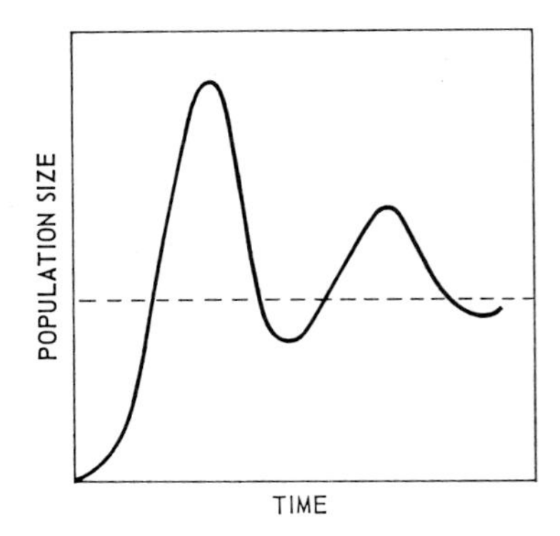

Figure 5. Population growth when the fertility and survival rates of an individual depend on the size of the population, both currently and at the time it was born. The broken line shows the size the population will be when stationarity is reached. (Redrawn from Leslie, 1959.)

4

Population Growth with Age-Dependent Birth and Death Rates II: The Continuous Time Model

1. Lotka's Equations

We have already shown (page 14) that, given a simple birth and death process in which the birth and death rates are unaffected either by population size or by the ages of the individuals, the growth of a population is described by the equation $N_t = N_0 e^{rt}$, or, alternatively, $N_{t+1} = N_t e^r$. Here N_t is the size of the population at time t, and r, the intrinsic rate of natural increase, is the difference between the instantaneous birth and death rates per head of the population. To conform with established usage these rates are, in this chapter, denoted by b and d, respectively, rather than by λ and μ. The instantaneous birth rate is defined as $b = B_t/N_t$, where B_t is the number of births occurring per unit of time at time t in the whole population of size N_t. It must be emphasized that B_t is a *rate*; it is necessarily defined in terms of a suitable unit of time, but it varies continuously; analogously, an accelerating automobile may at a single moment have an instantaneous speed of 50 mph. Further, whereas B_t is the rate for the population as a whole, b is the rate per individual. The instantaneous death rate d is defined similarly.

It was also shown (Chapter 3) that in a population in which the birth and death rates are age-dependent a stable age distribution may eventually be reached, and when this has happened

$$N_{t+1} = \lambda N_t,$$

where λ is the rate of growth per unit of time or the *finite rate of natural increase* in a population having a stable age distribution. It is evident that when the age distribution is stable the birth and death rates per head of the population as a whole remain constant; for although each age class has its own age-specific birth and death rates the proportions in which the age classes are present remain constant. So in these circumstances the two equations describing population growth are equivalent, and consequently

$$e^r = \lambda \quad \text{or} \quad \ln \lambda = r.$$

We now consider Lotka's original approach to the study of population growth. Time is treated as continuous and matrix methods are not used. What we wish to do is to estimate b and d (hence $r = b - d$) from observable quantities. Since two unknowns are to be derived, two equations that express b and d in terms of the observed quantities are necessary. As observables we take l_x and m_x, defined as follows:

l_x is the proportion of survivors at time x of an original cohort of individuals all born simultaneously at time 0.

m_x is the mean number of offspring born in unit time to an individual whose age is in the range $(x, x + dx)$.

It should be noticed that l_x is a pure number and m_x, a rate. Moreover, the rate m_x pertains only to those individuals in an infinitesimally small age class.

Two other quantities, to be eliminated in arriving at the final equations, are also required. These are

$c_x\, dx$, the proportion of the total population in the age range $(x, x + dx)$ when the age distribution is stable. The function c_x thus defines the stable age distribution.

$B_t\, dx$, the number of individuals born to the whole population, in an interval of length dx at time t.

Assuming the age distribution to be stable, it now follows that $N_t c_x\, dx$ is the size, at time t, of that fraction of the total population which consists of individuals whose ages are in the range $(x, x + dx)$. But these individuals are evidently the survivors of the $B_{t-x}\, dx$ individuals born at time $t - x$ in an interval of length dx, and thus

$$N_t c_x\, dx = l_x B_{t-x}\, dx. \tag{4.1}$$

Now, since the age distribution is stable, b, the instantaneous birth rate per head, is constant and equal to B_i/N_i for all i. Therefore

$$\begin{aligned} B_{t-x} &= bN_{t-x} \\ &= bN_t e^{-rx}, \end{aligned}$$

since

$$N_t = N_{t-x} e^{rx}.$$

Then, from (4.1),

$$c_x = \frac{l_x B_{t-x}}{N_t} = l_x b e^{-rx}.$$

But

$$\int_0^\infty c_x \, dx = 1$$

and so

$$\frac{1}{b} = \int_0^\infty l_x e^{-rx} \, dx. \tag{4.2}$$

This is the first of the two equations relating b and $d = b - r$.

Next suppose that the $N_t c_x \, dx$ individuals in the age range $(x, x + dx)$ and alive at time t produce m_x offspring per head per unit of time. Then the total number of newborn produced per unit time by all the population's age groups taken together is

$$B_t = \int_0^\infty N_t c_x m_x \, dx = \int_0^\infty b N_t l_x m_x e^{-rx} \, dx.$$

Therefore

$$\frac{1}{b} \frac{B_t}{N_t} = \int_0^\infty l_x m_x e^{-rx} \, dx.$$

But $b = B_t/N_t$ and so

$$\int_0^\infty l_x m_x e^{-rx} \, dx = 1. \tag{4.3}$$

This is the second of the required equations. When values of l_x and m_x have been found by experiment, (4.2) and (4.3) may be solved to yield b, d, and r. These equations cannot, unfortunately, be solved explicitly, and in practice it is necessary to replace the integrals with sums and obtain approximate solutions. Detailed accounts of appropriate methods will be found in papers by for instance, Leslie (1945), Birch (1948) and Leslie and Park (1949).

2. The Net and Gross Reproductive Rates and the Mean Generation Time

We have now mentioned r, the intrinsic rate of natural increase, and $\lambda = \text{antilog}_e\, r$, the finite rate of natural increase. Yet another function used to measure population growth is R_0, the net reproductive rate. This is the ratio between total births in two successive generations or the rate of growth per generation; that is

$$R_0 = \int_0^\infty l_x m_x \, dx.$$

As a function of T, the mean duration of a generation,

$$R_0 = \frac{N_{t+T}}{N_t} = e^{rT}.$$

This is equivalent to asserting that if we start with a cohort of size N it will, at a sequence of epochs separated by the fixed interval T, attain the sizes $R_0 N$, $R_0^2 N$, $R_0^3 N$, ..., although it is not assumed that in any one generation all the births occur simultaneously.

It is seen that $T = (\ln R_0)/r$. An approximate expression for the mean generation time T, when r is very small, is given by

$$T \doteq \frac{\int_0^\infty x l_x m_x \, dx}{\int_0^\infty l_x m_x \, dx}. \tag{4.4}$$

The rather intricate proof, given by Dublin and Lotka (1925), is as follows: First write

$$R_n = \int_0^\infty x^n l_x m_x \, dx. \tag{4.5}$$

We already have, by definition, $R_0 = e^{rT}$. To arrive at (4.4) it is necessary to find an expression of the form

$$R_0 = \exp[rf(R_0, R_1)].$$

The desired result will then be obtained by equating coefficients of r in the exponential terms. From (4.3)

$$1 = \int_0^\infty e^{-rx} l_x m_x \, dx.$$

Expanding the exponential in series and neglecting terms in r^2, r^3, ..., this gives

$$1 \doteq \int_0^\infty l_x m_x \, dx - r \int_0^\infty x l_x m_x \, dx.$$

Now put

$$y = \int_0^\infty e^{-rx} l_x m_x \, dx.$$

Then

$$\frac{dy}{dr} = -\int_0^\infty xe^{-rx} l_x m_x \, dx \equiv -Ay, \text{ say,}$$

where

$$A = \frac{\int_0^\infty xe^{-rx} l_x m_x \, dx}{\int_0^\infty e^{-rx} l_x m_x \, dx}.$$

Integrating the equation $dy/dr = -Ay$ yields

$$y = y_0 \exp\left(-\int A \, dr\right),$$

where

$$y_0 = \int_0^\infty l_x m_x \, dx = R_0$$

by definition. Hence

$$y = R_0 \exp\left(-\int A \, dr\right)$$

and, since $y = 1$, it follows that

$$R_0 = \exp\left(\int A \, dr\right). \tag{4.6}$$

Next A, hence $\int A \, dr$, must be evaluated. Writing both the numerator and the denominator of A in series form and using (4.5), we see that

$$A = \frac{R_1 - rR_2 + r^2 R_3/2! - r^3 R_4/3! + \cdots}{R_0 - rR_1 + r^2 R_2/2! - r^3 R_3/3! + \cdots}.$$

Thus A is a polynomial in r that may be written

$$A \equiv \alpha + \beta r + \gamma r^2 + \delta r^3 + \cdots,$$

where

$$\alpha = \frac{R_1}{R_0}, \qquad \beta = \alpha^2 - \left(\frac{R_2}{R_0}\right), \cdots.$$

Then, using (4.6), we see that

$$\ln R_0 = \int A \, dr = \alpha r + \frac{\beta r^2}{2} + \frac{\gamma r^3}{3} + \cdots \tag{4.7}$$

and

$$R_0 = \exp\left(\alpha r + \frac{\beta r^2}{2} + \frac{\gamma r^3}{3} + \cdots\right).$$

Notice that since the coefficients α, β, ..., are functions of the R_n we now have an equation relating r and the R_n only. Now recall that $R_0 = e^{rT}$. Thus

$$T = \alpha + \frac{\beta r}{2} + \frac{\gamma r^2}{3} + \cdots.$$

If r is so small that we may put $T = \alpha$ approximately, we then have

$$T \sim \frac{R_1}{R_0} = \frac{\int_0^\infty x l_x m_x \, dx}{\int_0^\infty l_x m_x \, dx}, \quad \text{as required.}$$

One may also use (4.7) to determine r if values of R_0, R_1, R_2, ..., have been obtained by experiment. Dublin and Lotka (1925), studying human populations with low r, were able to neglect terms beyond $\beta r^2/2$ in solving (4.7). Leslie and Ranson (1940), in determining the value of r for laboratory populations of the vole *Microtus agrestis*, found it necessary to use terms up to $\delta r^4/4$. Computational details will be found in their paper and also in Leslie (1945).

As a final measure of population growth—more precisely of "potential" growth—we have the "Gross Reproductive Rate" $G = \int_0^\infty m_x \, dx$. This is the expected number of offspring that would be born to an individual that survived for the whole of its reproductive period.

3. Age-Specific Fertility Tables

Before going further it must be emphasized that Lotka's equations will hold only in an environment sufficiently stable for the l_x and m_x figures to remain constant in time. Strictly, therefore, tables of l_x and m_x values should be compiled only from the results of controlled experiments throughout which environmental conditions are kept constant. Values of r, b, and d (and also, of course, of R_0, G, and T) derived from such tables apply only to populations living in the stated conditions.

Moreover, we have assumed in the foregoing that all the individuals in a population were capable of reproduction; that is, only the female part of a bisexual population was considered. The results are easily extended to cover a whole population (both sexes combined), provided the sex ratio within each age class remains constant.

The m_x values are variously called "age-specific fertility rates," "age-specific fecundity rates," or "maternal frequencies." The word "fertility" usually applies to species that bear live young. "Fecundity" is used for egg-laying species when what is counted is the numbers of eggs laid by females of different ages; in this case survival is measured from the moment an egg is laid.

Although m_x has been defined as the mean number of offspring born per unit time to a parent whose age is in the range $(x, x + dx)$, in practice parental age may be quite coarsely grouped. Thus we may take m_x equal to the mean number of offspring born per unit time to a parent in the age range $(x, x + h)$ and, provided h is small enough, the approximation will be satisfactory. This amounts to treating m_x as a step function instead of as a continuous variable.

It should be observed that m_x is not the same as the element F_x in the matrix $\mathbf{M}$ of Chapter 3 (page 34). The difference is this: m_x is the number of offspring born to a parent aged x in unit time and its value is unaffected by the subsequent fate of these offspring; F_x is that fraction of these m_x offspring that survive into the next time interval when time is treated as discrete. So to obtain F_x values, besides a table of m_x values, we also need a table of l_x values (survival rates). The derivation of F_x values from m_x and l_x tables is described in Leslie (1945).

4. Life Tables

The l_x values are normally portrayed in a "life table" in which, besides a column of the l_x values themselves, there are usually several other columns giving values derived from l_x; l_x is defined as the proportion of a cohort of individuals born at time 0 that still survive at time x when all are of age x. It is called the "life table function" or the "age-specific survival rate."

A life table for any species is of great interest by itself, even if we do not wish to estimate the population growth rate and the stable age distribution. From a life table we may infer at what age the risk of death is greatest; if we know the principal causes of death affecting the different age classes, we can infer to which of these causes most of the deaths are due, that is, which of the many dangers a species faces is most important in keeping its numbers down. Life tables are much studied by economic entomologists. For insect species that have nonoverlapping generations and whose eggs all hatch at roughly the same time a natural population is itself a cohort, and l_x values may be estimated directly by obtaining sample estimates of a population's size at a sequence of times. The form of the survivorship curve (a plot of l_x versus x) may suggest when a pest species is most likely to respond to control measures and also when control is most needed. For a pest that does damage during its immature stages (e.g., caterpillars) early destruction is desirable and it is worth knowing whether natural deaths occur predominantly early or late. For species that are pests in the adult stage (e.g., mosquitoes, blackflies, tsetse flies) it does not matter whether control measures are applied early or late in the immature stages so long as they succeed; thus the time to attempt control may depend on when, as judged from a life table, a species is most vulnerable.

A full "cohort life table" is prepared as follows: suppose we begin with a population of newborn individuals numbering l_0 (which may be set equal to 1, though often, to avoid small fractions, it is set equal to 1000). At a succession of times separated by equal intervals of one time unit a count is made of the number still alive. The results are tabulated as follows:

Age interval	Number of survivors at age x	Number of deaths in $(x, x+1)$	Proportion of deaths in $(x, x+1)$	Number of time units lived in $(x, x+1)$	Life remaining to those aged x	Observed expectation of life at age x
⋮	⋮	⋮	⋮	⋮	⋮	⋮
$(x, x+1)$	l_x	d_x	q_x	L_x	T_x	e_x
⋮	⋮	⋮	⋮	⋮	⋮	⋮

Clearly, for any $x > 0$, l_x is an observed value of a random variate, for, if the observations were repeated on another cohort, we should not expect to obtain identical l_x values. Because of this, l_x should always be called the *proportion* of survivors of age x, not the *probability* of being a survivor at age x. In the same way q_x is an observed proportion of deaths and e_x, an observed life expectation at age x. Indeed, all the entries in an empirical life table are to be thought of as sample values, not population values. We shall not deal here with their stochastic properties; the subject has been explored by Chiang (1960a, b).

The various columns of the table are interrelated as follows. Obviously

$$d_x = l_x - l_{x+1}$$

and

$$q_x = \frac{d_x}{l_x} = 1 - \frac{l_{x+1}}{l_x}.$$

The length of time lived during the interval $(x, x+1)$ by all individuals taken together is

$$L_x = \int_x^{x+1} l_x \, dx.$$

The units in which L_x is measured have the dimension individuals $\times$ time (analogous to "man-hours," say). Each individual alive at time x will contribute part of a unit if it dies during $(x, x+1)$ or one unit if it survives throughout the whole interval and is still alive at $x+1$. If the survivorship curve can be treated as approximately linear in $(x, x+1)$, we may write

$$L_x \doteq \tfrac{1}{2}(l_x + l_{x+1}).$$

The total lifetime remaining to all those members of the population alive at age x is the sum of their individual lifetimes; that is,

$$T_x = \sum_{j=x}^{w} L_x ,$$

where w is the age at which all are dead. Therefore, putting $L_x \doteq \frac{1}{2}(l_x + l_{x+1})$,

$$T_x \doteq \frac{l_x}{2} + \sum_{j=x+1}^{w} l_j .$$

The element P_x in the matrix $\mathbf{M}$ used to predict population growth when time is treated as discrete (see page 34) is then

$$P_x = \frac{L_{x+1}}{L_x} \doteq \frac{l_{x+1} + l_{x+2}}{l_x + l_{x+1}} \qquad \text{(see Leslie, 1945).}$$

Finally, e_x is the observed expectation of life for an individual aged x and is given by

$$e_x = \frac{T_x}{l_x} ;$$

that is, e_x is the total lifetime remaining at time x to the whole of the surviving population, divided by l_x, the number of individuals by whom it is shared.

The life table described above is known as a cohort life table. The l_x values for such a table can be obtained by direct observation only when the species concerned has a comparatively short lifespan and also when the fate of an even-aged population (i.e., a cohort) of the species can be followed until all its members are dead. This may be done with populations reared in the laboratory and it is sometimes possible with natural populations. However, for the latter not only must a natural cohort exist but we must also be able to recognise its members. Unless these requirements are met, life tables have to be derived indirectly. Caughley (1966, 1967) has described various methods of obtaining life-table functions of wild mammal populations.

It is also worthwhile to consider the methods used by demographers studying human populations. In this case what is observed is the existing age distribution in an all-age population and the number of deaths that occur in each age class during a single time interval. From these data a "current life table" is compiled. If we can safely assume that the data come from a stationary population, the values for a cohort life table may then be derived. It will be recalled (see page 37) that a population is "stationary" when (a) the age distribution is stable and (b) the total size of the population remains constant. A current life table is as follows (see Chiang 1960b). All entries refer to events occurring in the single unit of time for which the records are made.

Age class	Number of deaths in the age class	Total population at midpoint of time interval	Age-specific death rate
⋮	⋮	⋮	⋮
$(x, x+1)$	D_x	Q_x	$M_x = \dfrac{D_x}{Q_x}$
⋮	⋮	⋮	⋮

If, and only if, the population is stationary, D_x is the same as d_x in the cohort life table; that is

$$D_x = l_x - l_{x+1}.$$

Also

$$Q_x = L_x = \int_x^{x+1} l_x \, dx \doteq \tfrac{1}{2}(l_x + l_{x+1}),$$

the approximation being satisfactory, provided the time interval employed is short enough.

Then the age-specific death rate is

$$M_x = \frac{D_x}{Q_x} = \frac{l_x - l_{x+1}}{L_x} \doteq \frac{2(l_x - l_{x+1})}{l_x + l_{x+1}}.$$

Let us find the relationship between M_x, the age specific death rate and q_x, the proportion of deaths in the age class $(x, x+1)$. By definition,

$$q_x = 1 - \frac{l_{x+1}}{l_x}$$

and therefore

$$M_x = \frac{2q_x}{2 - q_x}$$

or

$$q_x = \frac{M_x}{1 + \frac{1}{2}M_x}.$$

Notice that q_x, which is an estimate of the probability that an individual will die when its age is in the range $(x, x+1)$, is obtained directly from a cohort life table but only indirectly from a current life table. It will be seen that the age specific death rate $M_x = D_x/Q_x$ is necessarily greater than $q_x = D_x/l_x$, since $l_{x+1} < l_x$ and therefore $Q_x \doteq \frac{1}{2}(l_x + l_{x+1}) < l_x$.

5

The Growth of Populations of Two Competing Species

1. The Deterministic Case

In Chapter 2 the logistic equation, $dN/dt = (a - bN)N$, was discussed. It will be recalled that this equation may in certain circumstances describe the growth of a single-species population of stable age distribution and living in a constant environment when resources are limited. The constants a and b are known as the logistic parameters of the species in the particular environmental conditions. It was assumed that only one species of organism was present and that the population of this one species would become as large as the available resources permitted.

We now extend the discussion by considering populations of two species living together and competing with each other for the same limiting resource. Each population is inhibited not only by members of its own species but also by those of the other. Denoting by N_i the number of individuals of species i (for $i = 1, 2$), it is reasonable to suppose that

$$\frac{dN_1}{dt} = N_1(a_1 - b_{11}N_1 - b_{12}N_2)$$

and (5.1)

$$\frac{dN_2}{dt} = N_2(a_2 - b_{21}N_1 - b_{22}N_2).$$

Here a_1 and b_{11} are the logistic parameters for species 1 if it is living alone; a_2 and b_{22} are the corresponding parameters for species 2. Also b_{12} measures the degree to which the presence of species 2 affects the growth of species 1 and vice versa for b_{21}.

In general, the simultaneous differential equations (5.1) cannot be explicitly solved. We consider first a particular set of circumstances in which they can be solved and return to the more general case afterward. The simplification that permits solution of the equations consists in assuming that for an individual of either species the inhibitory effect of all other individuals (of both species) is the same. Then each individual of both species behaves as if it were competing with a population of size $N = N_1 + pN_2$. The factor p allows for the fact that the members of the two species may differ from one another in their inhibitory effect. If individuals of species 2 make smaller inroads on the resources than individuals of species 1, then $p < 1$; conversely, if species 1 is the less demanding competitor, $p > 1$. In any case, N, the size of the "effective inhibiting population" is the same for both species.

Making this assumption, (5.1) may be replaced by

$$\frac{dN_1}{dt} = a_1 N_1 - (b_1 N)N_1$$
$$\frac{dN_2}{dt} = a_2 N_2 - (b_2 N)N_2. \tag{5.2}$$

It then follows that

$$b_{11}N_1 + b_{12}N_2 = b_1 N = b_1(N_1 + pN_2)$$

and

$$b_{21}N_1 + b_{22}N_2 = b_2 N = b_2(N_1 + pN_2).$$

Therefore

$$(b_{11} - b_1)N_1 + (b_{12} - b_1 p)N_2 = 0$$

and

$$(b_{21} - b_2)N_1 + (b_{22} - b_2 p)N_2 = 0,$$

so that

$$b_1 = b_{11}; \qquad \frac{b_{12}}{b_{11}} = p,$$
$$b_2 = b_{21}; \qquad \frac{b_{22}}{b_{21}} = p,$$

or

$$b_{12}b_{21} = b_{11}b_{22}.$$

This relation is therefore equivalent to the assumption already made, namely that the effective inhibiting population is of the same size for each of the two species. Equations 5.2 can now be rewritten

$$\frac{1}{N_1}\frac{dN_1}{dt} = \frac{d}{dt}\ln N_1 = a_1 - b_1 N$$

and

$$\frac{1}{N_2}\frac{dN_2}{dt} = \frac{d}{dt}\ln N_2 = a_2 - b_2 N.$$

Eliminating N gives

$$\frac{d}{dt}(b_2 \ln N_1 - b_1 \ln N_2) = \frac{d}{dt} \ln \left(\frac{N_1^{b_2}}{N_2^{b_1}}\right) = a_1 b_2 - a_2 b_1,$$

whence

$$\frac{N_1^{b_2}}{N_2^{b_1}} = \frac{[N_1(0)]^{b_2}}{[N_2(0)]^{b_1}} \exp[(a_1 b_2 - a_2 b_1)t],$$

where $N_i(0)$ for $i = 1, 2$, are the initial sizes of the two populations at $t = 0$.

Clearly, if the combined population is maintained long enough, only one of the species will persist and the other will die out. The outcome will depend only on whether $a_1 b_2$ is greater or less than $a_2 b_1$ and will be unaffected by the initial species numbers. If $a_1 b_2 > a_2 b_1$, ultimately species 1 will "win" and species 2, die out. The opposite will happen if $a_1 b_2 < a_2 b_1$. Once the losing species has become extinct, the winner will continue to grow in accordance with the single-species logistic process.

We now drop the simplifying assumption that each species is inhibited by the same effective population. Equivalently, we assert that $b_{12} b_{21} \neq b_{11} b_{22}$. There are four possible sets of circumstances to consider and these are most clearly shown graphically (see Figure 6). In the graphs the composition of the combined two-species population at any instant may be represented by a point having as coordinates the values of N_1 and N_2 at that instant.

Consider now the locus of points for which $dN_1/dt = 0$. This locus is the line (shown solid in the graphs)

$$a_1 - b_{11} N_1 - b_{12} N_2 = 0 \quad \text{or} \quad N_1 = \frac{a_1 - b_{12} N_2}{b_{11}}.$$

It cuts the N_1-axis at $N_1 = a_1/b_{11}$ and the N_2-axis at $N_2 = a_1/b_{12}$. If, at any instant, the point representing the combined populations falls below this line, that is, if

$$N_1 < \frac{a_1 - b_{12} N_2}{b_{11}},$$

then

$$\frac{dN_1}{dt} = N_1(a_1 - b_{11} N_1 - b_{12} N_2)$$

$$> N_1 \left[a_1 - b_{11} \cdot \frac{a_1 - b_{12} N_2}{b_{11}} - b_{12} N_2\right] = 0$$

and the species 1 population increases in size. Conversely, if

$$N_1 > \frac{a_1 - b_{12} N_2}{b_{11}}, \quad \text{then} \quad \frac{dN_1}{dt} < 0$$

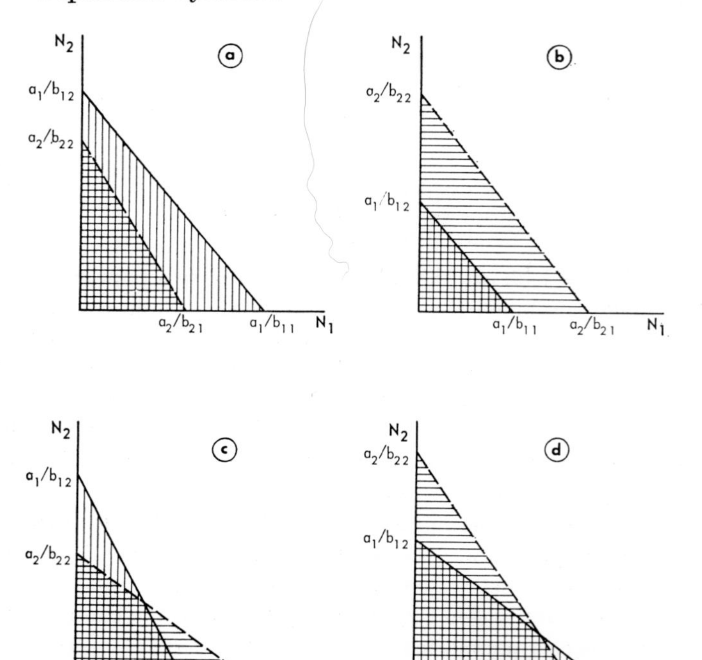

Figure 6. The four possible relationships between the lines $N_1 = (a_1 - b_{12}N_2)/b_{11}$, for which $dN_1/dt = 0$ (solid lines), and $N_2 = (a_2 - b_{21}N_1)/b_{22}$ for which $dN_2/dt = 0$ (dashed lines). In vertically hatched regions species 1 increases. In horizontally hatched regions species 2 increases. (*a*) Species 1 wins; (*b*) species 2 wins; (*c*) stable equilibrium; (*d*) unstable equilibrium.

and the species 1 population decreases. In the graphs the region in which species 1 will increase is shown by vertical hatching and the region in which this species will decrease, by the absence of vertical hatching.

Analogous arguments show that the line (dashed in the graphs)

$$a_2 - b_{21}N_1 - b_{22}N_2 = 0$$

represents the locus of points for which $dN_2/dt = 0$. It cuts the N_1-axis at $N_1 = a_2/b_{21}$ and the N_2-axis at $N_2 = a_2/b_{22}$. Whenever the point representing the combined populations falls below this line (region with horizontal hatching) $dN_2/dt > 0$ and the species 2 population increases in size. Conversely, above this line (horizontal hatching absent) species 2 must decrease.

Beginning with a combined population for which both N_1 and N_2 are small, so that the point representing the initial species composition falls within the cross-hatched area on any of the graphs, we now see that there are four possibilities, depending on the relative positions of the lines $dN_1/dt = 0$ and $dN_2/dt = 0$, hence on the values of the parameters. Thus, if, as in Figure 6*a*, the line $dN_1/dt = 0$ lies wholly above the line $dN_2/dt = 0$ or, in other words, if $a_1b_{22} > a_2b_{12}$ and $a_1b_{21} > a_2b_{11}$, then, although species 1 is sometimes capable of increasing when species 2 is stationary or decreasing, the reverse is never true. It follows that ultimately species 1 must replace species 2 entirely.

The opposite situation, in which species 2 will win and species 1 die out, is shown in Figure 6*b* for which $a_2b_{12} > a_1b_{22}$ and $a_2b_{11} > a_1b_{21}$.

In these two cases there exists no pair of values (N_1, N_2), with N_1 and N_2 both positive, that satisfies the simultaneous equations $dN_1/dt = dN_2/dt = 0$. An equilibrium condition in which the sizes of both populations remain constant is therefore impossible.

Now consider Figures 6*c* and 6*d* in which the lines $dN_1/dt = 0$ and $dN_2/dt = 0$ do cross each other. They intersect at the point

$$N_1 = \frac{a_1b_{22} - a_2b_{12}}{b_{11}b_{22} - b_{12}b_{21}} \quad \text{and} \quad N_2 = \frac{a_2b_{11} - a_1b_{21}}{b_{11}b_{22} - b_{12}b_{21}}.$$

For convenience let the coordinates of this point be denoted by $N_1(\text{eq})$ and $N_2(\text{eq})$. Theoretically, a combined population with precisely this composition would maintain itself unchanged indefinitely. Notice, however, that in one case the equilibrium is stable and in the other, unstable.

Figure 6*c*, in which $a_1b_{22} > a_2b_{12}$ and $a_2b_{11} > a_1b_{21}$, shows the conditions for stable equilibrium. Regardless of the initial numbers of the two species, as time passes they will steadily approach $N_1(\text{eq})$ for species 1 and $N_2(\text{eq})$ for species 2. This follows, since in the vertically hatched region in which $dN_1/dt \geq 0$ and $dN_2/dt \leq 0$ simultaneously $N_1 \leq N_1(\text{eq})$ and $N_2 \geq N_2(\text{eq})$. Correspondingly, in the horizontally hatched region in which $dN_1/dt \leq 0$ and $dN_2/dt \geq 0$ simultaneously $N_1 \geq N_1(\text{eq})$ and $N_2 \leq N_2(\text{eq})$. The situation is one of negative feedback.

The conditions for unstable equilibrium, namely, $a_2b_{12} > a_1b_{22}$ and $a_1b_{21} > a_2b_{11}$ are shown in Figure 6*d*. Here we have

$$\left.\begin{aligned} \frac{dN_1}{dt} &\leq 0 \\ \frac{dN_2}{dt} &\geq 0 \end{aligned}\right\} \text{(horizontally hatched), where } \begin{cases} N_1 \leq N_1(\text{eq}) \\ N_2 \geq N_2(\text{eq}) \end{cases},$$

and conversely

$$\left.\begin{aligned}\frac{dN_1}{dt} &\geq 0\\ \frac{dN_2}{dt} &\leq 0\end{aligned}\right\} \text{(vertically hatched), where } \begin{cases} N_1 \geq N_1(\text{eq}) \\ N_2 \leq N_2(\text{eq}) \end{cases}.$$

Consequently, any combined population must reach a state in which species 1 dies out and species 2 is left in sole possession of the resources or vice versa. Positive feedback occurs and therefore the equilibrium state is unstable. Whether species 1 or species 2 will be the winner depends not only on the values of the parameters but also on the sizes of the initial populations of the two species; to determine the outcome in any particular case we must trace the changes occurring in the composition of the combined populations as time passes.

Slobodkin (1961) has shown, with drawings of three-dimensional surfaces, how such two-species populations will behave in the various circumstances.

2. Determining the Trajectories of Two-Species Populations

It has already been remarked that the simultaneous differential equations (5.1) cannot, in general, be solved explicitly. To determine the course of events in a two-species population governed by these equations, therefore, it is necessary to find difference equations that permit prediction of the sizes of the two populations at time $t + 1$, given their sizes at time t; that is, we require expressions of the form

$$N_i(t+1) = f_i[N_1(t), N_2(t)] \quad \text{for} \quad i = 1, 2.$$

Then, given $N_1(0)$ and $N_2(0)$, we can calculate successively $N_1(1)$, $N_2(1)$, $N_1(2)$, $N_2(2)$, and so on. When the conditions are such that there is unstable equilibrium we may by this method discover which of the two competing species will be the winner for given values of $N_1(0)$ and $N_2(0)$.

It was shown on page 22 that simple logistic growth in a one-species population can be described either by the differential equation $dN/dt = N(a - bN)$ or by the difference equation

$$N(t+1) = \frac{\lambda N(t)}{1 + \alpha N(t)},$$

where $\lambda = e^a$ and $\alpha = b(\lambda - 1)/a$. This suggests that the differential equations (5.1) may be replaced by the difference equations

$$N_1(t+1) = \frac{\lambda_1 N_1(t)}{1 + \alpha_1 N_1(t) + \gamma_1 N_2(t)},$$

$$N_2(t+1) = \frac{\lambda_2 N_2(t)}{1 + \alpha_2 N_2(t) + \gamma_2 N_1(t)} \tag{5.3}$$

(see Leslie, 1958).

We shall now show that (5.3) is equivalent to (5.1), provided both α_i and γ_i are proportional to $\lambda_i - 1$. Also we shall find the parameters of the difference equations in terms of those of the differential equations. Only the first equation in (5.3) need be discussed, since identical arguments apply to the second equation.

Notice first that if species 1 were entirely unaffected by the presence of other members of *either* species we should have

$$N_1(t+1) = \lambda_1 N_1(t); \quad N_1(t+2) = \lambda_1^2 N_1(t); \ \ldots; \ N_1(t+h) = \lambda_1^h N_1(t).$$

Thus λ_1 is a finite rate of natural increase. Consequently, it is a function of the time interval concerned. In writing down an expression for $N_1(t+h)$ that accords with (5.3) we must therefore put λ_1^h in place of λ_1 wherever λ_1 occurs.

Writing $\alpha_1 = c(\lambda_1 - 1)$ and $\gamma_1 = c'(\lambda_1 - 1)$, where c and c' are constants of proportionality, it follows that

$$N_1(t+h) = \frac{\lambda_1^h N_1(t)}{1 + c(\lambda_1^h - 1)N_1(t) + c'(\lambda_1^h - 1)N_2(t)}.$$

Then

$$\frac{N_1(t+h) - N_1(t)}{h} = N_1(t)\,\frac{(\lambda_1^h - 1)}{h}\left[\frac{1 - cN_1(t) - c'N_2(t)}{1 + c(\lambda_1^h - 1)N_1(t) + c'(\lambda_1^h - 1)N_2(t)}\right].$$

When $h \to 0$, the left-hand side tends to dN_1/dt. Using l'Hôpital's rule, we see that

$$\lim_{h \to 0} \frac{\lambda_1^h - 1}{h} = \lim_{h \to 0} \frac{\lambda_1^h \ln \lambda_1}{1} = \ln \lambda_1.$$

Also note that, as $h \to 0$, the denominator of the expression in square brackets tends to 1. Therefore

$$\frac{dN_1}{dt} = N_1(t) \ln \lambda_1 [1 - cN_1(t) - c'N_2(t)].$$

This is equivalent to the first equation in (5.1) if

(i) $\ln \lambda_1 = a_1$, whence $\lambda_1 = e^{a_1}$,

(ii) $c \ln \lambda_1 = b_{11}$, whence $\alpha_1 = \dfrac{b_{11}(e^{a_1} - 1)}{a_1}$,

(iii) $c' \ln \lambda_1 = b_{12}$, whence $\gamma_1 = \dfrac{b_{12}(e^{a_1} - 1)}{a_1}$.

By the same arguments it may be shown that for the second equation in (5.3)

$$\lambda_2 = e^{a_2}, \qquad \alpha_2 = \frac{b_{22}(e^{a_2} - 1)}{a_2}, \quad \text{and} \quad \gamma_2 = \frac{b_{21}(e^{a_2} - 1)}{a_2}.$$

Thus, if the constants in (5.1) are given, those in (5.3) may be obtained. Then for any preassigned composition of the initial population, say $[N_1(0), N_2(0)]$, the sequence $[N_1(t), N_2(t)]$ for $t = 1, 2, \ldots,$ may be calculated and the trajectory of the combined population plotted. The use of different starting points leads to a whole family of such trajectories. Examples are given in Figures 7 and 8.

3. Stochastic Simulation of Population Growth with Two Competing Species

In the deterministic case described above it is necessary to know only the parameters of (5.1) and, if they are such that they give unstable equilibrium, the numbers of each species initially to be able to predict the outcome of competition with certainty. For natural populations in which births and deaths are to some extent matters of chance this is not so. There is always a risk that a species that had been expected to succeed, either alone or in company with its competitor, will disappear because of an unlucky preponderance of deaths over births for a limited period.

It is therefore interesting to generate stochastic simulations of the competition process from which we can estimate empirically the probability that the theoretically expected result will actually happen in given conditions. Before simulating the process, it is necessary to write (5.1) in more detail, remembering that they represent only a net balance of births and deaths. As in the simulation of one-species logistic growth (see page 22), so also in this case, it is necessary to treat birth rates and death rates separately. Thus, since the equation

$$\frac{dN_1}{dt} = (a_1 - b_{11}N_1 - b_{12}N_2)N_1$$

represents a net result of births and deaths, we write it more fully in the form

$$\frac{dN_1}{dt} = N_1[\lambda_1(N_1 \mid N_2) - \mu_1(N_1 \mid N_2)],$$

where $\lambda_1(N_1 \mid N_2)$ is the birth rate of species 1 in the presence of N_1 members of its own species and N_2 members of the other and $\mu_1(N_1 \mid N_2)$ is the death rate in the same circumstances. Then, if

$$\lambda_1(N_1 \mid N_2) = a_{1b} - b_{11b}N_1 - b_{12b}N_2$$

and

$$\mu_1(N_1 \mid N_2) = a_{1d} + b_{11d} N_1 + b_{12d} N_2,$$

we have

$$\frac{dN_1}{dt} = N_1[(a_{1b} - a_{1d}) - (b_{11b} - b_{11d})N_1 - (b_{12b} - b_{12d})N_2]$$

Similarly,

$$\frac{dN_2}{dt} = N_2[(a_{2b} - a_{2d}) - (b_{21b} - b_{21d})N_1 - (b_{22b} - b_{22d})N_2].$$

These equations take into account the fact that both the birth rate and the death rate of each species are affected by the number of members of its own species and the number of members of the competing species present at any time.

The possible events in the population and their probabilities are given in Table 3, where the constant C is such that the probabilities sum to unity.

TABLE 3

EVENT	PROBABILITY
Species 1 birth: $N_1 \to N_1 + 1$; $N_2 \to N_2$	$CN_1\lambda_1(N_1 \mid N_2)$
Species 1 death: $N_1 \to N_1 - 1$; $N_2 \to N_2$	$CN_1\mu_1(N_1 \mid N_2)$
Species 2 birth: $N_1 \to N_1$; $N_2 \to N_2 + 1$	$CN_2\lambda_2(N_2 \mid N_1)$
Species 2 death: $N_1 \to N_1$; $N_2 \to N_2 - 1$	$CN_2\mu_2(N_2 \mid N_1)$

Simulation of the intervals between successive events can also be done if desired by the method described on page 24. Usually, however, all that is needed is a plot of N_2 versus N_1 and the time intervals are of less interest. Barnett (1962) gives a number of examples of computer simulations of the fate of two competing species. Besides plots of N_2 versus N_1, he also states for each example the number of steps (i.e., births and deaths) that occurred before one of the species became extinct.

4. Competing Species and Competitive Exclusion

A discussion of the coexistence of two competing species would not be complete without some mention of the bearing of the mathematical arguments in this chapter on the well-known coexistence principle or competitive exclusion principle of ecology. The extensive literature on the subject is not

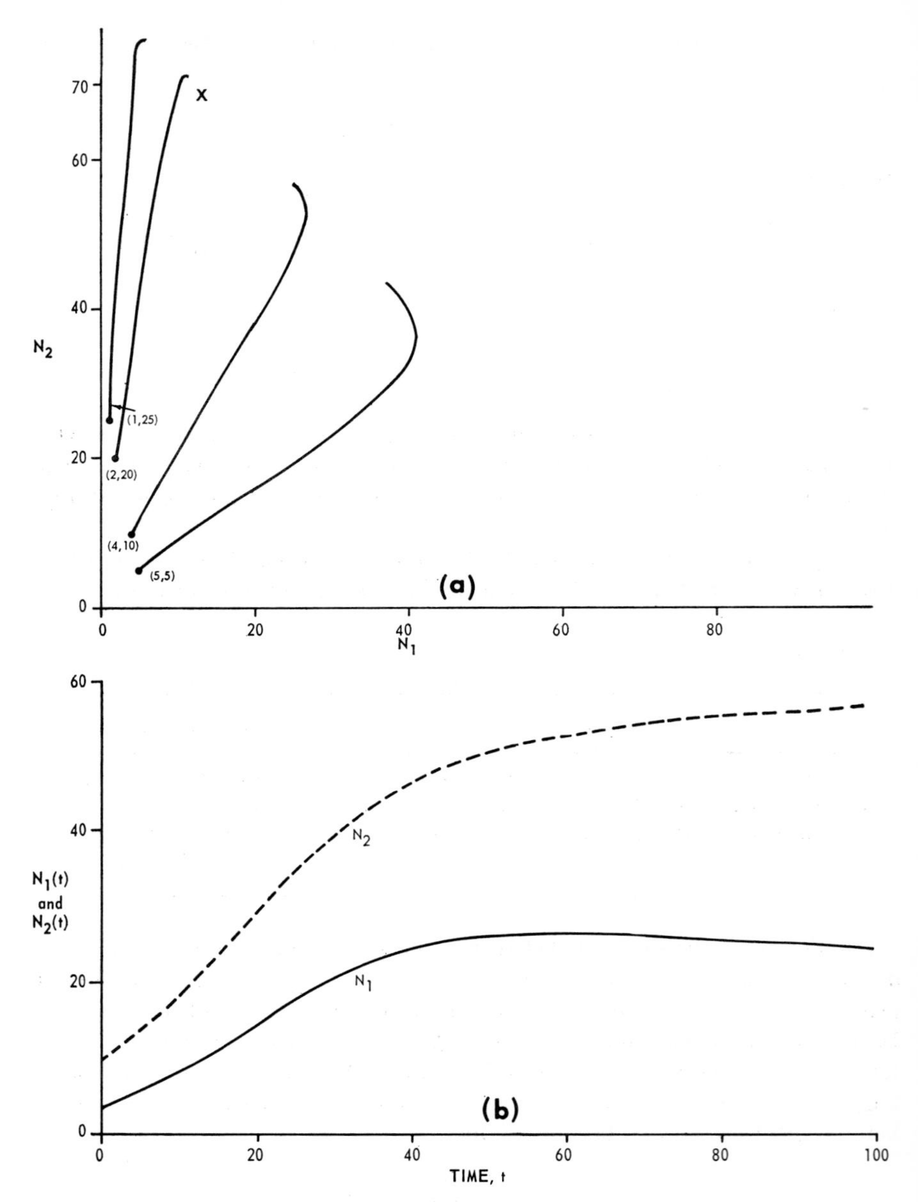

Figure 7*a*. Four curves of the family

$$dN_1/dt = N_1(0.1 - 0.0014N_1 - 0.0012N_2),$$
$$dN_2/dt = N_2(0.08 - 0.0009N_1 - 0.001N_2).$$

The stable equilibrium point is at $N_1 = 12.5$, $N_2 = 68.75$. The sizes of the initial populations $[N_1(0), N_2(0)]$ are shown at the start of each curve. All curves are shown as far as $t = 100$.

Figure 7*b*. The variation with time of N_1 and N_2, given the starting conditions $N_1(0) = 4$, $N_2(0) = 10$, up to $t = 100$. Ultimately they will level off with $N_1 = 12.5$ and $N_2 = 68.75$.

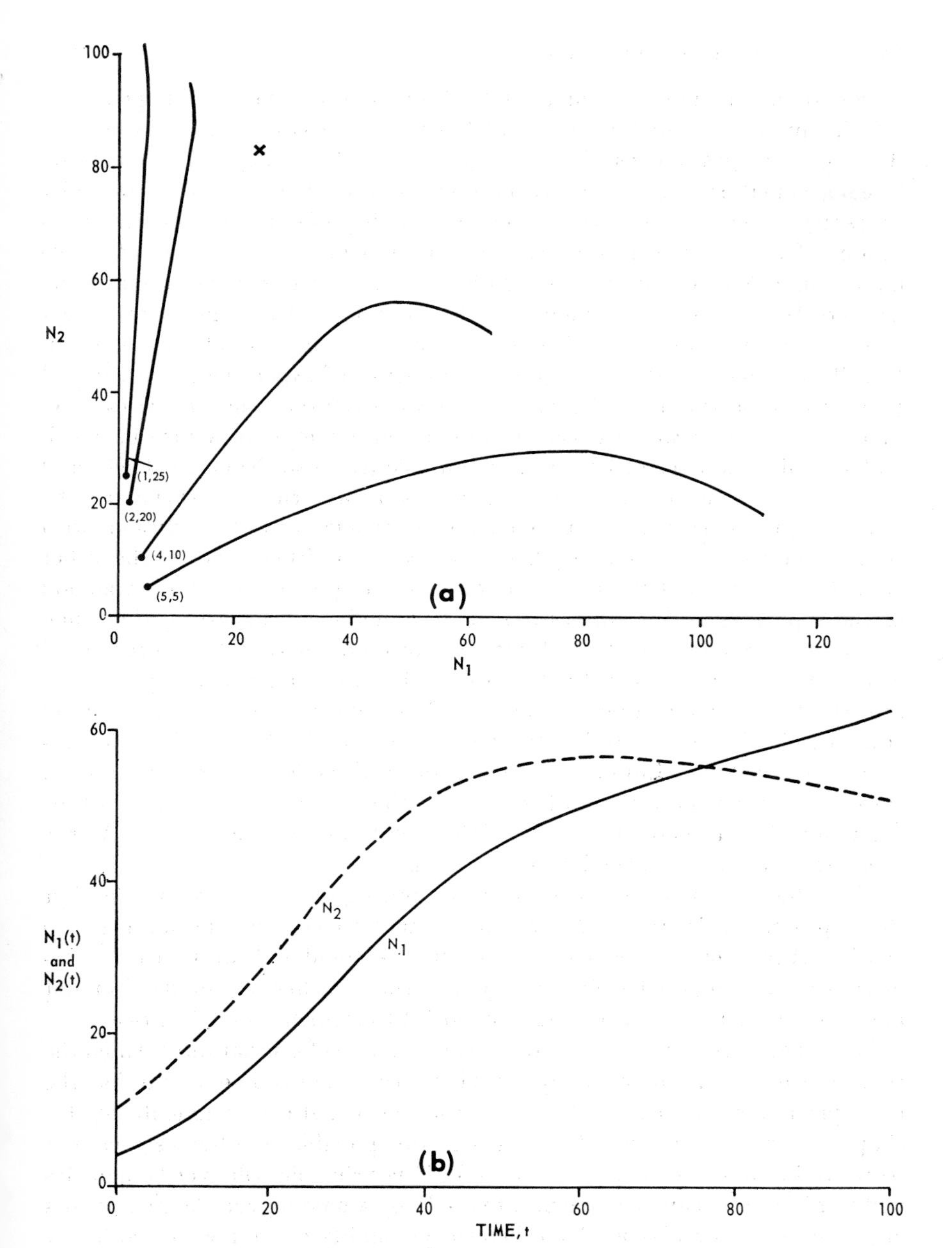

Figure 8*a*. Four curves of the family

$$dN_1/dt = N_1(0.1 - 0.0007N_1 - 0.001N_2),$$
$$dN_2/dt = N_2(0.075 - 0.0007N_1 - 0.0007N_2).$$

The equilibrium point (unstable) is at $N_1 = 23.8$, $N_2 = 83.3$. The sizes of the initial populations are shown at the start of each curve. All curves are shown as far as $t = 100$.

Figure 8*b*. The variation with time of N_1 and N_2 given $N_1(0) = 4$, $N_2(0) = 10$, up to $t = 100$. Ultimately species 2 will become extinct.

reviewed here; it suffices to quote DeBach's (1966) statement of the principle: "Different species which co-exist indefinitely in the same habitat must have different ecological niches; that is, they must not be ecological homologues." The mathematical arguments advanced above show that this is not a logically necessary conclusion. We have shown that it is perfectly possible for populations of two competing species to coexist indefinitely in a state of stable equilibrium. No stipulations were made concerning whether the species were, or were not, ecological homologues, but the fact that, theoretically, coexistence of ecological homologues can happen does not, of course, mean that it does happen. Park's (1954) famous series of experiments with mixed populations of the flour beetles *Tribolium confusum* and *T. castaneum*, maintained in laboratory cultures, provides an example of a case in which stable equilibrium seemed to be unattainable (see also, Neyman, Park, and Scott, 1956). One or the other species was always the sole survivor. The mixed populations were maintained under various sets of conditions with several replicates of each set. Under certain conditions one or the other species won consistently. In other conditions *T. confusum* won in a proportion of the replicates and *T. castaneum* in the remainder, suggesting that such conditions brought about unstable equilibrium, making the outcome of competition unpredictable. In these cases, therefore, it seems likely that the growth of the combined populations could be described, at least roughly, by equations in the form of (5.1) with $a_2 b_{12} > a_1 b_{22}$ and $a_1 b_{21} > a_2 b_{11}$, the conditions for unstable equilibrium. The theoretical model is certainly inexact, since it makes no allowance for time lags: birth and death rates at any instant are assumed to depend on the sizes of the populations at that instant. All the same, it may provide a good approximation.

These flour beetles are not ecological homologues, as shown by the fact that optimum cultural conditions were found to be different for the two species when each was cultured by itself. We could well argue that if coexistence were impossible for two species whose niches are similar but not identical the same would be true a fortiori if their niches *were* identical.

The experiments therefore constitute an example of a situation in which the coexistence principle seems to apply. This does not, of course, prove that the principle is true in general. Since, as the mathematical arguments of this chapter have shown, coexistence is logically possible, it obviously cannot be proved impossible. Thus no "proof" of the principle will ever be had. Its truth, if it is true, can be demonstrated only by amassing very large amounts of evidence, all in its favor. Even then the possibility of finding contradictory evidence will always remain.

Returning to the mathematical argument, it is interesting to compare the equilibrium sizes of two populations, when they are competing, with the size each would attain if it were growing alone. In the absence of competitors the saturation level for a population of the ith species is a_i/b_{ii} ($i=1, 2$). Recalling that (see page 57)

$$N_1(\text{eq}) = \frac{a_1 b_{22} - a_2 b_{12}}{b_{11} b_{22} - b_{12} b_{21}}, \qquad N_2(\text{eq}) = \frac{a_2 b_{11} - a_1 b_{21}}{b_{11} b_{22} - b_{12} b_{21}},$$

we see that the condition for stable equilibrium is equivalent to $N_i(\text{eq}) < a_i/b_{ii}$. This is what we should expect. Each population stops increasing before it reaches such a size that it is using all the resources; thus some are left for the competing species.

6

The Dynamics of Host-Parasite Populations

1. The Lotka-Volterra Equations

In the preceding chapter we considered populations of two species in which, owing to competition, each species inhibited the multiplication of the other. An entirely different form of interaction between two species occurs when one species is a parasite (or predator) and the other its host (or prey). The two possibilities, "host-parasite" and "predator-prey," are mathematically equivalent, and in what follows we speak of hosts and parasites. Clearly, the more abundant the host (H), the most opportunities there are for the parasite (P) to breed. However, as the parasite population grows, the number of hosts destroyed by parasites increases. So the simplest pair of equations pertaining to possible population growth is that originally described by Lotka and Volterra (cf. Lotka, 1925), namely,

$$\frac{dH}{dt} = (a_1 - b_1 P)H,$$
$$\frac{dP}{dt} = (-a_2 + b_2 H)P, \tag{6.1}$$

with a_1, a_2, b_1, $b_2 > 0$. These equations are, of course, wholly deterministic and make no allowance for stochastic fluctuations. The parameter a_1 is the net growth rate per individual of the host species in the absence of the parasite; this rate of increase is diminished by an amount $b_1 P$ when P parasites are present. The parasite population, on the other hand, dwindles to nothing in the absence of hosts, since reproduction is then impossible; parasite increase is directly proportional to the number of hosts present.

To solve this pair of equations, note that

$$\frac{dH}{dP} = \frac{(a_1 - b_1 P)H}{(-a_2 + b_2 H)P}$$

or

$$a_2 \frac{dH}{H} - b_2\,dH + a_1 \frac{dP}{P} - b_1\,dP = 0.$$

On integration we then have

$$a_2 \ln H - b_2 H + a_1 \ln P - b_1 P = \text{constant.} \tag{6.2}$$

Equation 6.2 represents a family of closed curves in which each member of the family corresponds to a different value of the constant. Choice of a starting point, that is, of initial values of H and P, determines the constant. Three such curves, all with the same parameter values, are shown in Figure 9. Any population will continue indefinitely to follow the cyclical path on which it starts, in an anticlockwise direction. There is no damping toward the equilibrium point at the center of the curves; this is the point $H = a_2/b_2$, $P = a_1/b_1$ at which $dH/dt = dP/dt = 0$.

When the cycles are of very small amplitude, they may be approximated by ellipses. It is worthwhile to derive the equation of such an ellipse, since some interesting properties of the process can be inferred from it. To show that small cycles are nearly elliptical, we proceed as follows (see Chiang, 1954).

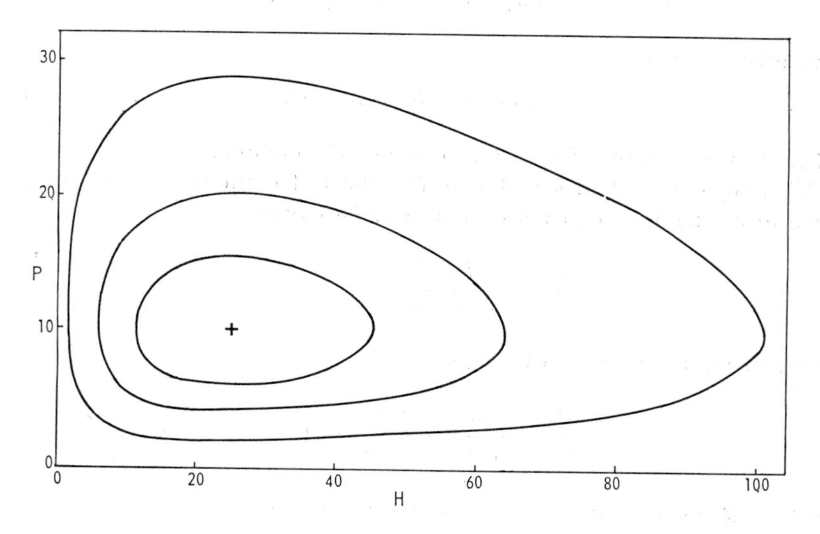

Figure 9. Three curves of the family $dH/dt = (a_1 - b_1 P)H$; $dP/dt = (-a_2 + b_2 H)P$, with $a_1 = 1.00$, $b_1 = 0.10$, $a_2 = 0\ 50$ $b_2 = 0.02$. The equilibrium point is $H = 25$, $P = 10$.

First shift the origin of the coordinates to the equilibrium point (a_2/b_2, a_1/b_1) and in the new frame write h and p for the numbers of hosts and parasites, respectively. Thus

$$h = H - \frac{a_2}{b_2}, \qquad p = P - \frac{a_1}{b_1}.$$

Then (6.1) becomes

$$\frac{dh}{dt} = a_1\left(h + \frac{a_2}{b_2}\right) - b_1\left(h + \frac{a_2}{b_2}\right)\left(p + \frac{a_1}{b_1}\right) = \frac{-b_1 a_2}{b_2} p - b_1 hp$$

and (6.3)

$$\frac{dp}{dt} = b_2\left(h + \frac{a_2}{b_2}\right)\left(p + \frac{a_1}{b_1}\right) - a_2\left(p + \frac{a_1}{b_1}\right) = \frac{a_1 b_2}{b_1} h + b_2 ph.$$

In the neighborhood of the equilibrium point (now the origin) the product ph becomes negligibly small and terms involving it may be disregarded. Then

$$\frac{dh}{dp} = -\frac{p}{h}\frac{b_1{}^2 a_2}{b_2{}^2 a_1}$$

or

$$b_2{}^2 a_1 h\, dh + b_1{}^2 a_2 p\, dp = 0.$$

Integrating, we have

$$b_2{}^2 a_1 h^2 + b_1{}^2 a_2 p^2 = C, \tag{6.4}$$

where C is a constant. This is the equation of an ellipse.

We now wish to find h and p as functions of t, the time. Neglecting the term in ph the first equation in (6.3) may be written

$$\frac{dt}{dh} = -\frac{b_2}{b_1 a_2}\frac{1}{p}.$$

Substituting for p from (6.4) yields

$$\frac{dt}{dh} = \frac{-b_2}{b_1 a_2}\left(\frac{b_1{}^2 a_2}{C - b_2{}^2 a_1 h^2}\right)^{1/2}.$$

Rewriting this in a form easy to integrate, we have

$$\sqrt{a_1 a_2}\, dt = \frac{-dh}{\sqrt{(C/b_2{}^2 a_1 - h^2)}},$$

which on integration yields

$$\sqrt{a_1 a_2}\, t + \beta = \cos^{-1}\left[\frac{h}{\sqrt{C/b_2{}^2 a_1}}\right]$$

or

$$h = \frac{\sqrt{C}}{b_2\sqrt{a_1}} \cos(\sqrt{a_1 a_2}\, t + \beta).$$

The constant of integration β is in the range $[0, 2\pi]$. Now put $\sqrt{C} = \sqrt{a_1 a_2}\, \alpha$; then $h = (a_2/b_2)\alpha \cos(\sqrt{a_1 a_2}\, t + \beta)$. Finally, returning to the original coordinate frame, we have

$$H = \frac{a_2}{b_2} + \frac{a_2}{b_2} \cdot \alpha \cos(\sqrt{a_1 a_2}\, t + \beta).$$

In the same way it may be shown that

$$P = \frac{a_1}{b_1} + \frac{a_1}{b_1}\left(\frac{a_2}{a_1}\right)^{1/2} \cdot \alpha \sin(\sqrt{a_1 a_2}\, t + \beta).$$

From these equations four properties that are true of small cycles (in the neighborhood of the equilibrium point) follow:

1. The sizes of both the host and the parasite populations vary sinusoidally with period $T = 2\pi/\sqrt{a_1 a_2}$. The period is the same for both species and depends only on the parameters a_1 and a_2.

2. The two populations are always one $\frac{1}{4}$-cycle out of phase. Thus the host population begins to decrease from its maximum size at the instant when the parasite population is undergoing its fastest growth; one $\frac{1}{4}$-cycle later, when the host population is declining most rapidly, the parasite population attains its maximum size and starts to decline. The magnitude of the constant of integration β depends only on which instant of time is chosen to represent $t = 0$. If we set $t = 0$, when H is a maximum, then $\beta = 0$.

3. The amplitudes of the oscillations are $\alpha a_2/b_2 = (1/b_2)(\sqrt{C/a_1})$ for the host population and $(\alpha a_1/b_1)\sqrt{a_2/a_1} = (1/b_1)(\sqrt{C/a_2})$ for the parasite population. They therefore depend on the constant of integration C introduced in (6.4) as well as on the parameters; that is, the amplitudes, unlike the period, depend on the initial sizes of the populations.

4. The mean size of the host population over a period T is

$$\frac{1}{T}\int_{t_0}^{t_0+T} H\, dt = \frac{1}{T}\left[\frac{a_2 t}{b_2}\right]_{t_0}^{t_0+T} = \frac{a_2}{b_2};$$

this is identical with the equilibrium size of the host population. Correspondingly, the mean size of the parasite population is

$$\frac{1}{T}\int_{t_0}^{t_0+T} P\,dt = \frac{a_1}{b_1}.$$

2. Stochastic Simulation of Host-Parasite Population Growth

The stochastic version of the deterministic process described by the Lotka-Volterra equations can, of course, be simulated. The procedure is like that already described for simulating logistic growth of a one-species population (Chapter 2) or the growth of populations of two competing species (Chapter 5). As was true for the earlier cases, the process described by

$$\frac{dH}{dt} = (a_1 - b_1 P)H, \qquad \frac{dP}{dt} = (-a_2 + b_2 H)P,$$

with given parameter values, can arise from many different combinations of birth and death rates; the equations represent only the net outcome.

However, for simplicity, suppose that a_1 is a pure birth rate for the hosts and that all host deaths are due to parasite attack. Assume also that a_2 is a pure death rate for the parasites and that the birth of a parasite always coincides with the death of a host. This would be true, for instance, if the parasites were parasitic wasps that laid their eggs in the larvae of host insects so that development of each parasite necessarily involved destruction of a host larva (or pupa). In such a host-parasite population the number of possible events is either three or four. The case with three possible events arises when every death of a host is accompanied by the birth of a parasite; in this case $b_1 = b_2 = b$, say, and the events and their probabilities are shown in Table 4, where C is a normalizing constant.

TABLE 4

EVENT	PROBABILITY
$H \to H+1;\ P \to P$	Ca_1H
$H \to H;\ P \to P-1$	Ca_2P
$H \to H-1;\ P \to P+1$	$CbPH$

Now suppose that some host deaths do not coincide with parasite births. This would happen if some parasite attacks were unsuccessful so that, though the host was killed by the attack, no parasite was born as a result. Let a proportion θ of parasite attacks be successful so that $b_2 = \theta b_1$. Then the event $H \to H - 1$ will have two possible accompaniments: either $P \to P + 1$, with probability proportional to $\theta b_1 HP = b_2 HP$, or $P \to P$, with probability proportional to $(1 - \theta)b_1 HP = (b_1 - b_2)HP$. It will be seen that if $\theta = 1$ we shall have the three-possibilities case already described.

The risk of extinction of one or the other of the species is clearly very great when the deterministic path of the combined population (as shown by the closed curves in Figure 9) passes close to either of the axes. When a population reaches a point in its trajectory such that one of the species has only a few members, the probability that that species will fail to recover is great. If the parasite dies out, the host population will grow without restriction until density dependence comes into operation. If the host dies out, the parasite too must die out soon afterward, since it will be unable to reproduce. Bartlett (1957, 1960) gives an example of a simulated trajectory of the stochastic form of the Lotka-Volterra process.

3. Other Equations for Host-Parasite Population Growth

The Lotka-Volterra equations allow for the influence of parasites on hosts and of hosts on parasites, but they contain no terms to allow for *intra*specific interference. We might write

$$\frac{dH}{dt} = (a_1 - b_1 H - c_1 P)H, \qquad \frac{dP}{dt} = (-a_2 + c_2 H)P, \tag{6.5}$$

where it is assumed that the hosts (but not the parasites) interfere with one another; this leads to the term in H^2 in the first equation.

Another possibility arises when we are dealing with predator (P) and prey (H) populations if we assume that the predator can survive on alternative food, although the presence of its proper prey H favors its multiplication. Then

$$\begin{aligned} \frac{dH}{dt} &= (a_1 - b_1 H - c_1 P)H, \\ \frac{dP}{dt} &= (a_2 - b_2 P + c_2 H)P. \end{aligned} \tag{6.6}$$

Here a_1 and b_1 are the logistic parameters for the prey species existing alone; a_2 and b_2 are the corresponding parameters for the predator species in the absence of its favored prey. The term in HP is negative in the first equation, since predators reduce the prey population, and positive in the second equation, since an abundance of prey favors the predators.

The sets of equations (6.5) and (6.6) are not considered further here. Still another possibility, and one that we shall go into, has been described by Leslie and Gower (1960). Suppose

$$\frac{dH}{dt} = (a_1 - c_1 P)H, \qquad \frac{dP}{dt} = (a_2 - c_2 P/H)P. \tag{6.7}$$

The term in P/H in the second equation arises from the fact that this ratio must affect the growth of the parasite population. When parasites are numerous and hosts scarce, P/H is large, and growth of the parasite population slows down. Conversely, when the supply of hosts is ample for the parasites present, P/H is small, and there is but slight restriction on parasite multiplication.

From these differential equations we may obtain the difference equations that specify the state of affairs at time $t + 1$, given that at time t. By the same arguments that were used in Chapter 5 (see page 59) to show the equivalence of (5.3) and (5.1) it is seen that the first equation in (6.7) may be replaced by

$$H(t+1) = \frac{\lambda_1 H(t)}{1 + \gamma_1 P(t)},$$

where $\lambda_1 = e^{a_1}$ and $\gamma_1 = c_1(e^{a_1} - 1)/a_1$.

We show next that the second equation in (6.7) is equivalent to the difference equation

$$P(t+1) = \frac{\lambda_2 P(t)}{1 + \gamma_2 P(t)/H(t)},$$

where $\gamma_2 = k(\lambda_2 - 1)$, with k a constant. Recall that, in writing $P(t+h)$, $\lambda_2{}^h$ must be substituted for λ_2 (see page 59). Then

$$\frac{P(t+h) - P(t)}{h} = P(t)\,\frac{(\lambda_2{}^h - 1)}{h}\left[\frac{1 - kP(t)/H(t)}{1 + k(\lambda_2{}^h - 1)P(t)/H(t)}\right].$$

Thus, when $h \to 0$,

$$\frac{dP}{dt} = P(t) \ln \lambda_2 \left[1 - \frac{kP(t)}{H(t)}\right],$$

which is the same as $dP/dt = (a_2 - c_2 P/H)P$ if $\lambda_2 = e^{a_2}$ and $\gamma_2 = c_2(e^{a_2} - 1)/a_2$.

The trajectory of a combined population following this process differs markedly from that of a Lotka-Volterra population. In the Lotka-Volterra case a plot of H versus P gave an endlessly repeated curve of roughly elliptical shape. In the present case the curve of H versus P forms a closing spiral that converges on the equilibrium point (see Figure 10); this is the point

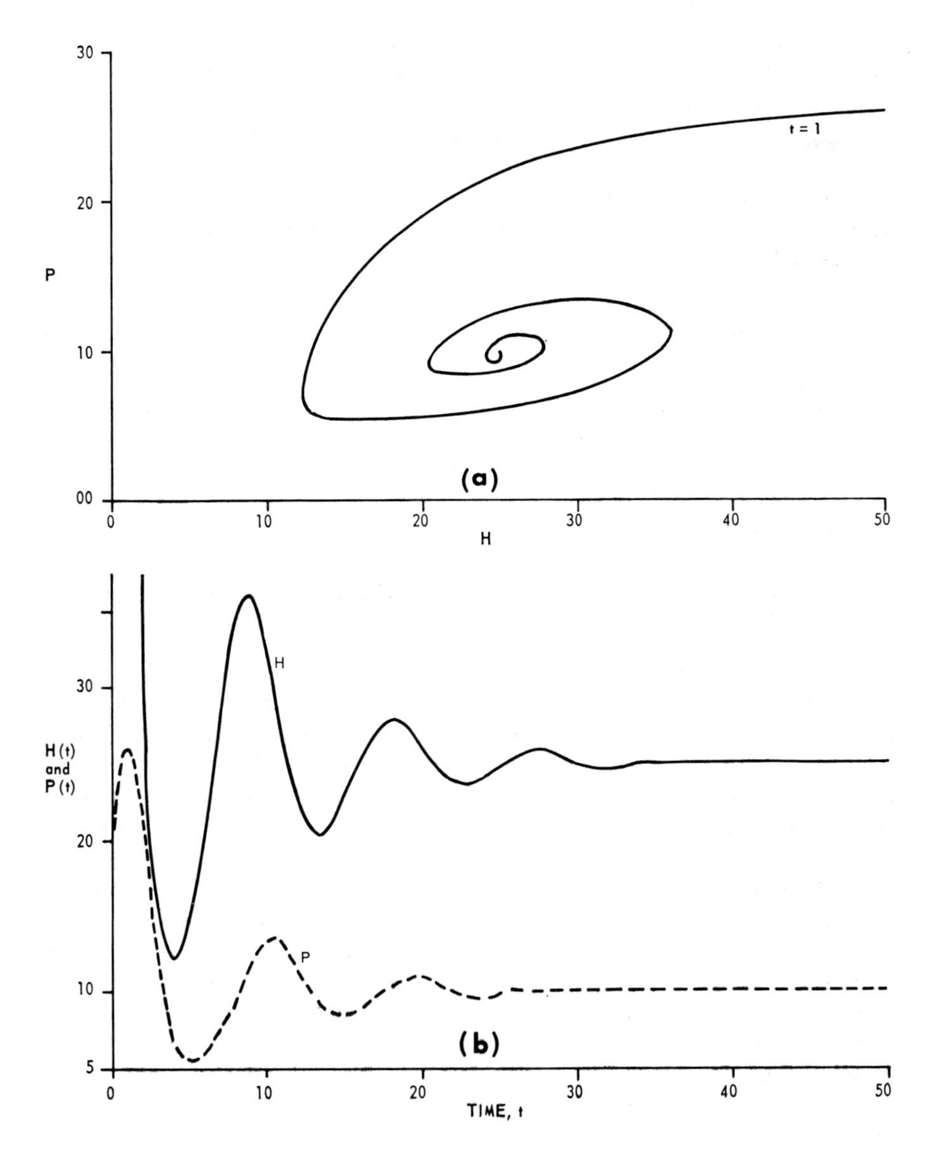

Figure 10*a*. An example of the curve $dH/dt = (a_1 - c_1P)H$; $dP/dt = (a_2 - c_2P/H)P$, with $a_1 = 1.0, c_1 = 0.1, a_2 = 1.0, c_2 = 2.5$. Initially, $H_0 = 80$, $P_0 = 20$. The equilibrium point is at $H = 25$, $P = 10$.
Figure 10*b*. The variation with time of H and P, given $H_0 = 80$, $P_0 = 20$.

$H = a_1c_2/a_2c_1$, $P = a_1/c_1$, where $dH/dt = dP/dt = 0$. Consider now the variation of H and P with respect to time: in the Lotka-Volterra case the oscillations persist with constant amplitude; in the present case the oscillations are damped so that their amplitudes steadily decrease until equilibrium is reached.

Slightly more complicated than (6.7) are the following equations discussed by Leslie and Gower (1960); the term in H^2 in the first of them allows for density dependence among the hosts.

$$\frac{dH}{dt} = (a_1 - b_1H - c_1P)H, \qquad \frac{dP}{dt} = (a_2 - c_2P/H)P.$$

The properties of this set of equations are similar to those of (6.7). Again, both H and P exhibit damped oscillations with time; or, what comes to the same thing, the H versus P trajectory is a spiral converging on the equilibrium point. In the stochastic case there is obviously less risk of either species becoming extinct, because the trajectory steadily draws away from the axes, where the chance of extinction is greatest.

4. General Remarks on Simple Models in Population Dynamics

The conceptual models of population change that we have considered in the first six chapters of this book have all be extremely simple. Although they are certainly too simple to be realistic, it is still worth knowing how populations governed by such rules will behave. This knowledge is a necessary preliminary to the development of more elaborate models. Among the many complicating factors we have neglected, but which are likely to affect real populations, are nonuniformity, both spatial and temporal, of the environmental conditions, individual differences among the organisms constituting a population, spatial clumping of the organisms so that the effects of density dependence will not be everywhere the same, immigration and emigration, and the effects of time lags in the responses of organisms to environmental change.

It is sometimes found that actual populations behave in a manner very similar to that of simple models in which no allowance is made for any of these complications. There are three possible explanations, not necessarily mutually exclusive.

1. The factors neglected may indeed be of negligible importance in some cases.
2. Some of the neglected factors may cancel one another out.
3. The resemblance of a model to the real-life process it is intended to represent may not be as close as it seems to be.

As remarked earlier (page 31), a superficial resemblance between model and reality may be found if we do no more than compare the outcome of the conceptual processes of a model and the natural processes of a living population. Quite different conceptual models may lead to fairly similar curves of population change. Then the data obtained by observing population sizes at a succession of times are not likely to be adequate for discriminating among the possible models.

Any model is bound to contain several assumptions, not just one, and ways of testing the assumptions one at a time must be sought. In studying the dynamics of laboratory reared populations, separate experimental tests of the different assumptions are sometimes possible. Smith's (1963) experiments on *Daphnia magna* populations, described in Chapter 2, are an example. Experiments with field populations are usually more difficult but can occasionally be done; for instance, Hairston (1967) succeeded in carrying out field experiments on populations of *Paramecium aurelia* in rainwater seepage ponds in a hardwood forest; he manipulated the environment both by supplying artificial rain and by augmenting the natural population with additional laboratory-rear Paramecia.

There are, of course, two quite distinct ways in which we may attempt to improve on an overly simple deterministic model. One is to make more realistic, and therefore more complicated, assumptions while still treating the process as deterministic. The other is to allow for the effect of stochastic variation, assuming a fairly simple model. Thus in Park's experiments with two competing species of *Tribolium* (see page 64) it will be recalled that identical replicates gave contrasting results. There was no need, however, to postulate subtle differences among the replicates and devise separate models allowing for the supposed differences. A single model in which stochastic variation was permitted sufficed to explain the inconsistent results.

The fact that repeated trials with the same simple stochastic model can lead to diametrically opposed outcomes should never be overlooked by those who devise ever more complicated deterministic models.

II

Spatial Patterns in One-Species Populations

7

Spatial Patterns and their Representation by Discrete Distributions

1. Introduction

In considering the way the size of a population of organisms fluctuates in time, as a result of births and deaths, we have sometimes taken density dependence into account; that is, we have assumed that birth rates and death rates must vary, depending on the size of the population already present. When we consider a population of sessile (or, at least, sedentary) organisms occupying a large area or volume, it is clear that unless all the organisms are evenly spaced out the density effect must vary from place to place. Where organisms are grouped in dense clusters, their mutual interference must be far greater than at sites in which they are sparse. Thus for so-called "patchily dispersed" organisms it is unreasonable to suppose that the birth and death rate per individual are functions of the number of members in the population as a whole. It is for this reason that the discussions in the preceding chapters on temporal changes in population sizes can be expected to correspond with reality only in populations of motile organisms contained within a space small enough for any organism to be likely to move freely throughout the whole of it. Only then would it be reasonable to suppose that all population members affect one another equally.

This brings us to the study of *spatial patterns* of populations of organisms. Much work has been done on the arrangements, or patterns, of organisms at one moment in time. Patterns need to be described and accounted for, and the task has proved so engrossing that so far little headway has been made in combining the two approaches to the study of populations: that of tracing how they vary with time and that of accounting for their patterns in space. These two topics are therefore still somewhat isolated from each other, and in embarking on a study of spatial patterns we make very little reference to results in population dynamics.

The change in topics entails a change in the sort of populations to be visualized. In discussing population dynamics, we were thinking of small, motile, fast-breeding organisms that occupy a fairly small volume with definite boundaries. It was necessary to postulate a small volume so that environmental conditions might be the same everywhere; the organisms themselves were small so that a small space could contain many of them; motility was taken for granted, thus ensuring that localized concentrations of individuals, if they ever occurred, would not persist; and rapid breeding was implicitly assumed so that we might reasonably suppose that conditions would remain constant over several generations or at least that any change would be too slight to be noticeable in the course of an experiment. For population dynamics is, in the main, an experimental subject. Although the conclusions arrived at may be extrapolated to some types of field population of the kind ecologists usually deal with, they are certainly more convincing if additional confirmation is obtained from laboratory-reared populations. Extrapolation to populations for which the assumptions are, strictly speaking, unreasonable is often illuminating if cautiously done, but in developing theoretical population dynamics small populations of small, fast-breeding organisms are what is usually visualized.

In any study of spatial patterns quite different populations must be thought of. We are now considering sessile or sedentary organisms whose pattern persists long enough to be worth studying. We are not greatly concerned with the duration of a generation or with birth and death rates; and the area (or volume) occupied by a population may be large, as may be the organisms it contains. Anything from the limpets on a rock to the trees in a forest may be involved. Moreover, although population dynamics is at present chiefly a branch of animal ecology, the study of spatial patterns is of equal concern to both plant and animal ecologists.

In considering the pattern, or dispersion, of a collection of organisms throughout a definite tract of space, it is necessary, first, to distinguish three wholly different setups that depend on the nature of the space and the organisms.

1. Cases in which the organisms are confined to discrete habitable sites or "units"; for example, caterpillars of a pest species that attack the shoots of a tree. Each shoot constitutes a habitable site and is a natural sampling unit. The caterpillars will not be found elsewhere than on shoots, and therefore the space available to them is discontinuous. Migrations from one shoot to another, even if possible, are assumed to be uncommon enough to be ignored. Thus, if we count the number of organisms per unit (in the example, caterpillars per shoot) for a large sample of the units, the observations clearly convey something about the spatial pattern of the species.

2. Cases in which the organisms have a continuum of space that they can occupy: for example, trees in a forest. There are now no natural sampling units such as the shoots of the preceding example. If we wish to count individuals per unit, the unit (a small plot of ground, or quadrat, for instance) has to be arbitrarily defined.

3. Cases in which, in addition to the absence of natural sampling units, there are no clearly delimited individuals that can be counted. This is a situation that often confronts plant ecologists. Not merely do plants occupy a continuum that cannot be subdivided naturally but also, owing to vegetative reproduction, a great many plant species do not occur as distinct individuals amenable to counting.

The problems raised by cases 2 and 3 are considered in later chapters. Chapters 7 and 8 deal with case 1, and we begin by supposing that we have available an observed frequency distribution of the number of individuals per unit for a sample of units selected at random from an extensive population. What will the observed frequency distribution be like and how can it be explained?

2. The Poisson Distribution

If the individuals are assigned independently and at random to the available units, we call their pattern (or dispersion) *random*, and expect to find that the number of individuals per unit is a Poisson variate. The probability that a unit will contain r individuals is then

$$p_r = \frac{\lambda^r e^{-\lambda}}{r!},$$

where λ is the mean number of individuals per unit.

To see this, suppose that every unit contains a large number, n, of locations, each of which can be occupied by a single individual. For every location, in every unit, the probability is the same, say p, that it will be occupied. Then the probability that exactly r locations in any one unit will be occupied is given by the binomial probability

$$\binom{n}{r} p^r (1-p)^{n-r}.$$

Now suppose n is very large, p is very small, and the mean of the distribution np is of moderate magnitude. Write $np = \lambda$. Also assume that r is negligibly small compared with n. Then

$$p_r = \binom{n}{r} p^r (1-p)^{n-r} \sim \frac{(np)^r}{r!} (1-p)^n$$

$$= \frac{\lambda^r (1-\lambda/n)^n}{r!} \to \frac{\lambda^r e^{-\lambda}}{r!}$$

as n becomes indefinitely large. In other words, when the individual organisms are sparse in relation to what the whole collection of units could contain and every possible location within a unit has the same probability of being occupied, the number of individuals per unit will be a Poisson variate.

This argument assumes that the maximum number of individuals a unit could contain is the same for all units and is equal to n and also that the expected number per unit is the same for all units and is equal to $np = \lambda$. These are very restrictive assumptions and are seldom likely to hold. It is not surprising therefore that the Poisson distribution rarely fits observed frequency distributions of the number of individuals per unit.

The probability generating function (pgf) of the Poisson distribution is given by

$$g(z) = p_0 + p_1 z + p_2 z^2 + \cdots$$

$$= e^{-\lambda} + \lambda e^{-\lambda} z + \frac{\lambda^2 e^{-\lambda}}{2!} z^2 + \cdots$$

$$= e^{\lambda(z-1)}.$$

Therefore the mean is

$$\left. \frac{dg(z)}{dz} \right|_{z=1} = g'(1) = \lambda$$

and the variance is

$$g''(1) + g'(1)[1 - g'(1)] = \lambda,$$

where $g''(1)$ denotes the value of $d^2 g(z)/dz^2$ at $z = 1$. Thus for the Poisson distribution the mean and variance are equal. However, when we compare the sample mean and variance of observed frequency distributions of the number of organisms per unit, it is commonly found that the variance greatly exceeds the mean. When this is so the pattern is said to be "aggregated," "clumped," "clustered," or "patchy" and the frequency distribution itself is described as "contagious." We now search for more realistic hypotheses to account for the distributions actually found. These fall into three categories: generalized distributions, compound distributions, and distributions with added zeros.

Before describing them, however, some comment on terminology is desirable. Much confusion exists in the ecological literature because the word "distribution" is used in both its colloquial and statistical senses, even sometimes in a single sentence, and without any explanation of the meaning intended. Colloquially, "distribution" is synonymous with "arrangement" or "pattern." Statistically, it means the way in which variate values are apportioned, with different frequencies, in a number of possible classes. In this sense there is no implied reference to spatial arrangement; for instance, we may speak of the distribution of a variate such as tree-height without any thought of the location of the trees. Thus to talk of a population of insects, say, as having "a clumped distribution with a large variance" is nonsense; it is the insects themselves that are clumped or have a clumped pattern, and the large variance pertains to the distribution of the variate, the number of insects per sample unit. To avoid ambiguity in statistical ecology it is most desirable to use the word distribution in its statistical sense *only*. Then a variate has a distribution, whereas a collection of organisms has a pattern.

3. Generalized Distributions

A generalized distribution arises if we suppose that groups or clusters of individuals (rather than single individuals) constitute the entities having a specified pattern and that the number of individuals per group is a random variate with its own probability distribution.

An example will illustrate this. Suppose female insects lay egg clusters that are dispersed at random among available units such as pine shoots. Denote the mean number of clusters per shoot by λ_1. The number of egg clusters per shoot will then have a Poisson distribution with pgf

$$G(z) = e^{\lambda_1(z-1)}.$$

Next, suppose that the number of larvae hatching from each cluster is itself a Poisson variate with mean λ_2. The pgf of the number of larvae per cluster is thus

$$g(z) = e^{\lambda_2(z-1)}.$$

The pgf of the number of individual larvae per shoot is then given by $H(z)$, where

$$H(z) = G(g(z)) = \exp[\lambda_1(e^{\lambda_2(z-1)} - 1)].$$

This is the pgf of the Neyman Type A or Poisson-Poisson distribution.

From the pgf we may obtain the distribution's mean and variance and its general term p_r, the probability that a unit will contain exactly r individuals. Thus the mean is $H'(1) = \lambda_1\lambda_2$. The variance is $H''(1) + H'(1)[1 - H'(1)] = \lambda_1\lambda_2(1 + \lambda_2)$, which obviously exceeds the mean.

The coefficient of z^r in the expansion of $H(z)$ gives p_r. Noting that

$$H(z) = e^{-\lambda_1} \exp[\lambda_1 e^{-\lambda_2} e^{\lambda_2 z}]$$

$$= e^{-\lambda_1} \left[1 + \lambda_1 e^{-\lambda_2} \sum_{j=0}^{\infty} \frac{(\lambda_2 z)^j}{j!} + \frac{(\lambda_1 e^{-\lambda_2})^2}{2!} \sum_{j=0}^{\infty} \frac{(2\lambda_2 z)^j}{j!} + \cdots \right],$$

we see that p_r, which is the coefficient of z^r, is

$$p_r = e^{-\lambda_1} \left[\lambda_1 e^{-\lambda_2} \frac{\lambda_2^r}{r!} + \frac{(\lambda_1 e^{-\lambda_2})^2}{2!} \frac{(2\lambda_2)^r}{r!} + \frac{(\lambda_1 e^{-\lambda_2})^3}{3!} \frac{(3\lambda_2)^r}{r!} + \cdots \right]$$

$$= e^{-\lambda_1} \frac{\lambda_2^r}{r!} \sum_{j=0}^{\infty} \frac{(\lambda_1 e^{-\lambda_2})^j}{j!} \cdot j^r.$$

A method of evaluating this function recursively will be found in Archibald (1948).

As a second example of a generalized distribution, we suppose, as before, that the clusters are randomly dispersed and that the pgf of the number of clusters per unit is

$$G(z) = e^{\lambda(z-1)}.$$

Now, instead of assuming that the number of larvae per cluster has a Poisson distribution, we let it have the logarithmic distribution with parameter α. This is equivalent to saying that $P(x)$, the probability that a cluster will contain x larvae, is proportional to α^x/x with $0 < \alpha < 1$. Here x takes the values 1, 2, ..., and no "empty" clusters, devoid of larvae, are postulated, as they were when x was assumed to be a Poisson variate.

Since

$$\sum_{x=1}^{\infty} P(x) = 1$$

and

$$\sum_{x=1}^{\infty} P(x) \propto \alpha + \frac{\alpha^2}{2} + \frac{\alpha^3}{3} + \cdots = -\ln(1-\alpha),$$

we see that

$$P(x) = \frac{-1}{\ln(1-\alpha)} \frac{\alpha^x}{x}.$$

The pgf of the logarithmic distribution is then

$$g(z) = \frac{-1}{\ln(1-\alpha)} \left[\alpha z + \frac{(\alpha z)^2}{2} + \frac{(\alpha z)^3}{3} + \cdots \right]$$

$$= \frac{\ln(1-\alpha z)}{\ln(1-\alpha)}.$$

For the generalized distribution, therefore, we obtain the pgf $H(z)$ by writing

$$H(z) = G(g(z)) = \exp\left\{\lambda\left[\frac{\ln(1-\alpha z)}{\ln(1-\alpha)} - 1\right]\right\}.$$

This is most easily simplified by redefining the two parameters λ and α as follows: Put $\lambda = k \ln Q$ and $\alpha = P/Q$, where $Q = 1 + P$. Now, in place of λ and α we have two other parameters, k and Q (or $P = Q - 1$). Then, since

$$\ln(1-\alpha) = \ln\frac{Q-P}{Q} = \ln\frac{1}{Q},$$

$$H(z) = \exp\left[-k\ln\left(1 - \frac{Pz}{Q}\right)\right] \cdot \exp[-k\ln Q]$$

$$= \left(1 - \frac{Pz}{Q}\right)^{-k} Q^{-k} = (Q - Pz)^{-k}.$$

This is the pgf of the negative binomial distribution. It may be contrasted with the pgf of the ordinary positive binomial for which $q, p < 1$, $p + q = 1$, and k is positive. For the negative binomial, on the other hand, the index is negative and $Q - P = 1$.

As before, we find the mean, variance, and general term p_r of the distribution. The mean is $H'(1) = kP$; the variance is $H''(1) + H'(1)[1 - H'(1)] = kP(1 + P)$, or kPQ. Instead of taking P and k as the parameters of the distribution, it is often preferable to take, as one parameter, the mean $m = kP$. The variance is then $V = m + m^2/k$. Clearly the smaller the value of k, the greater the variance; but, if $k \to \infty$, $V \to m$, and we see that, as for the Poisson distribution, the mean and variance are equal. Indeed, as $k \to \infty$, $p_r \to m^r e^{-m}/r!$, as proved below.

The probability p_r of finding exactly r individuals in a unit is the coefficient of z^r in the expansion of $H(z)$.

Now,

$$H(z) = Q^{-k}\left(1 - \frac{Pz}{Q}\right)^{-k}$$

$$= Q^{-k}\left[1 + \frac{kP}{Q}z + \frac{k(k+1)}{2!} \cdot \left(\frac{P}{Q}\right)^2 z^2 + \cdots\right].$$

Therefore the coefficient of z^r is

$$p_r = Q^{-k}\frac{k(k+1)\cdots(k+r-1)}{r!}\frac{P^r}{Q^r} = \frac{\Gamma(k+r)}{r!\,\Gamma(k)} \cdot \frac{P^r}{Q^{k+r}}$$

Recalling that the mean is $m = kP$, we may write instead

$$p_r = \frac{\Gamma(k+r)}{r!\,\Gamma(k)} \cdot \left(\frac{m}{k}\right)^r \left(\frac{k}{k+m}\right)^{k+r}$$

$$= \frac{\Gamma(k+r)}{\Gamma(k)k^r} \cdot \frac{m^r}{r!} \left(1 + \frac{m}{k}\right)^{-(k+r)}.$$

Then, as k becomes indefinitely large and r becomes negligible in comparison with k, $p_r \to m^r e^{-m}/r!$, which is the general term of the Poisson series.

We have now considered two examples of generalized distributions: the Neyman type A or Poisson-Poisson and the negative binomial or Poisson-logarithmic. In these double names for generalized distributions the first denotes the distribution of the number of clusters per unit and the second, the distribution of the number of individuals per cluster.

As an example of how these two distributions fit field observations, consider the data given by Bliss and Fisher (1953) in which both theoretical distributions were fitted to the observed distribution of the number per quadrat of *Salicornia stricta* plants growing in a salt marsh. Although in this case the sampling units are not discrete natural entities but arbitrary quadrats, the data serve for purposes of illustration. The way in which the parameters of the two theoretical distributions were estimated from the observations is described in the original paper by Bliss and Fisher and is not dealt with here, but, as can be seen from Figure 11, both theoretical distributions fit the data fairly well. By judging the goodness of fit with χ^2 tests it was found that for the negative binomial $P(\chi^2) = 0.48$ and for the Neyman type A, $P(\chi^2) = 0.17$. Thus, although the negative binomial appears to give a better fit, either of the hypotheses might be accepted. In other words, *if* we are willing to accept that the individuals occur as randomly dispersed clusters or clumps, the number of individuals per cluster may equally well be a Poisson variate or a logarithmic variate. A definite conclusion cannot be reached. Quite possibly, however, neither of these explanations accounts for the observed pattern. We now show how an entirely different hypothesis can lead to the negative binomial distribution.

4. Compound Distributions

Suppose that the organisms are independent of one another (or not clustered) and that if all the units available to them were identical their pattern would be random. If the mean density were λ, the probability that any unit would contain r individuals would therefore be the Poisson term $\lambda^r e^{-\lambda}/r!$.

Now suppose that the units are dissimilar. Some provide more favorable environments than others so that the parameter λ, the expected number of individuals in a unit, varies from unit to unit; that is, λ is itself a random

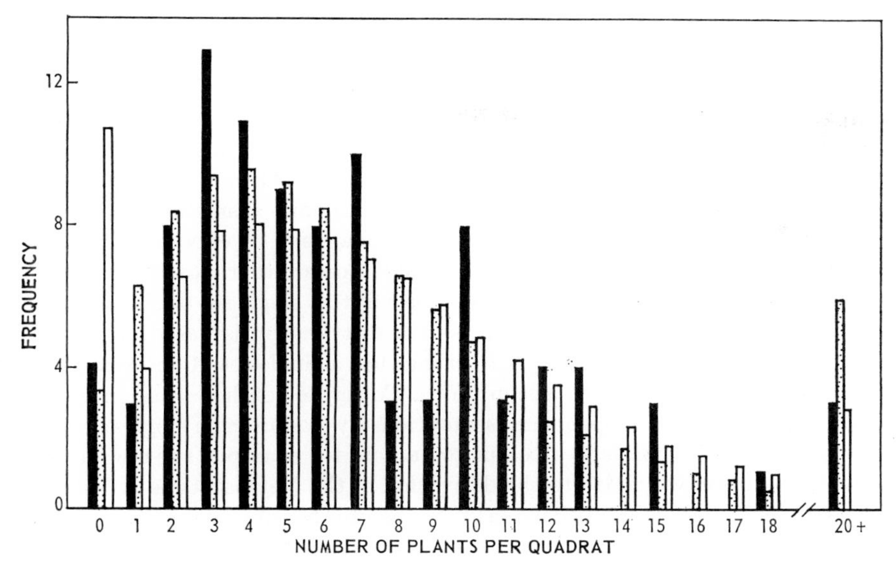

Figure 11. The distribution of the number of plants per quadrat of *Salicornia stricta*. Solid bars: observed. Stippled bars: fitted negative binomial distribution for which $P(\chi^2) = 0.48$. Open bars: fitted Neyman Type A distribution for which $P(\chi^2) = 0.17$. (Data from Bliss and Fisher, 1953.)

variable. Let us assume it has a Pearson type III distribution. The reason for choosing this standard curve to represent the distribution of λ is that whatever λ's true distribution it is likely that some type III curve can be found to approximate it closely. A true type III variate may have any non-negative value and the curve may be unimodal or J-shaped. Therefore we assume that the probability density function of λ is

$$f(\lambda) = \frac{1}{\Gamma(k)}\left(\frac{1}{P}\right)^k \lambda^{k-1}e^{-\lambda(1/P)}, \qquad (\lambda \geq 0).$$

Then

$$p_r = \frac{1}{r!\Gamma(k)}\left(\frac{1}{P}\right)^k \int_0^\infty \lambda^{r+k-1}e^{-\lambda(1+1/P)}\, d\lambda$$

$$= \frac{P^{-k}}{r!\Gamma(k)} \cdot \frac{\Gamma(r+k)}{\left(\dfrac{1+P}{P}\right)^{r+k}} = \frac{\Gamma(r+k)}{r!\Gamma(k)}\frac{P^r}{Q^{k+r}},$$

where $Q = 1 + P$. Notice that this is the distribution that we considered before—the negative binomial.

It is indeed true, in general, that every compound distribution corresponds to a generalized distribution and vice versa. A proof of this appears in Feller (1943). We have given only one well-known example here. Others have been described by Skellam (1952). Therefore it is futile to try to reach conclusions about the mechanism underlying a particular observed pattern simply by examining the observed distribution of the number of individuals per unit. Even when only one of the theoretical contagious distributions fits the observations there are two explanatory mechanisms to choose from. Also, since many of the theoretical series resemble one another closely, it is often found that two or more provide an adequate fit to a single set of observations.

Since every generalized, and compound, distribution is based on two (at least) assumptions, it is scarcely surprising that a single set of observations is inadequate to confirm both assumptions. To try to reach a decision on the acceptability of two independent assumptions by examining a single observed frequency distribution is to attempt too much; for example, the mechanism, described above, which led to the derivation of the Neyman type A distribution contained two assumptions: (a) that the number of clusters per unit was a Poisson variate and (b) that the number of individuals per cluster was also a Poisson variate. Only if one of these assumptions were believed, from *independent* evidence, to be true could the acceptability of the other be judged by fitting a Neyman type A series and seeing how good a fit was obtained.

It must be concluded that the fitting of theoretical frequency distributions to observational data can never by itself suffice to "explain" the pattern of a natural population.

5. Distributions with Added Zeros

Another mechanism that may produce aggregation is the following: the units sampled may fall into two classes that are not visibly distinguishable. Suppose that one class consists of units that for unknown reasons are uninhabitable and cannot be occupied by any individuals and that the other class consists of habitable units among which the individuals are dispersed at random. Then the variate, the number of individuals per unit, will have a variance greater than its mean and the distribution will be Poisson with added zeros.

If the proportion of habitable units is θ and of uninhabitable, $1 - \theta$, then

$$p_0 = 1 - \theta + \theta e^{-\lambda}$$

and

$$p_r = \frac{\theta \lambda^r e^{-\lambda}}{r!}, \qquad (r = 1, 2, \ldots .).$$

Notice that the empty units consist of a mixture of those that cannot be inhabited and those that, though habitable, chance to be empty.

For this distribution the mean is $\theta\lambda$ and the variance $\theta\lambda[1 + \lambda(1 - \theta)]$.

This is the simplest of the distributions with added zeros. It has only two parameters; θ, the proportion of habitable units, and λ, the Poisson parameter for these units. Distributions with more parameters can easily be derived by postulating that in those of the units that are habitable the number of individuals has one of the contagious distributions.

8

The Measurement of Aggregation

1. Introduction

It was shown in Chapter 7 that even when one of the theoretical contagious distributions fits an observed frequency distribution closely it is still not possible to draw any conclusions regarding the mechanism that gave rise to it. However, we may still wish to measure the degree of aggregation (clumping, clustering, or contagion) of a population's spatial pattern without attempting to explain it. It would then be possible to compare the aggregation exhibited by a single species at different times or in different places or to compare the aggregation found at a single place and time in populations of two different species. Such observations are of obvious ecological interest and various methods of measuring aggregation have been devised. It should be emphasized that we are still considering organisms that occur only in discrete habitable units (Case 1 on page 80).

Usually, two populations that are to be compared will differ in mean density as well as in degree of aggregation. Attempts have been made to define some measurable property of a population's spatial pattern that can be thought of as equivalent in some way to aggregation but independent of mean density. In fact, of course, we cannot really contemplate aggregation without at the same time thinking of the things that are aggregated, hence of their numbers. Indeed, the phrase "degree of aggregation" describes a vague, undefined notion that is open to several interpretations. If aggregation is to be measured, we must first choose from a number of possibilities some measurable property of a spatial pattern that is to be called its aggregation, and the method of measurement is then implicit in the chosen definition. Thus the several existing ways of measuring aggregation are not different methods of measuring the same thing: they measure different things. All the measures make use of the observed frequency distribution of the number of individuals per unit.

2. The Variance/Mean Ratio

The defining property of the contagious discrete distributions is that the variance exceeds the mean for all of them, whereas for the Poisson distribution the variance and mean are equal. This immediately suggests use of the variance: mean ratio V/m as a measure of aggregation. The sample value of this ratio is $(1/n\bar{x}) \sum_{j=1}^{n} (x_j - \bar{x})^2$, where x_j is the number of individuals in the jth of n units sampled and $\bar{x} = (\sum x_j)/n$. For large n the expected value of the ratio when the individuals are dispersed at random is $E(V/m) \sim 1$. If a population yields a value of V/m that exceeds 1 only slightly, we may be tempted to enquire whether the ratio exceeds 1 significantly or whether the value obtained would quite likely have been yielded by a randomly dispersed population. This is easily tested in virtue of the fact that $\sum(x_j - \bar{x})^2/\bar{x}$ (known as the index of dispersion) is the sum of n terms of the form $(O - E)^2/E$, where O and E are the observed and expected frequencies of the individuals in each unit. The sum is therefore approximately distributed as a χ^2 variate with $n - 1$ degrees of freedom. The number of degrees of freedom is one less than the number of variate values observed, since the values are subject to the constraint $\sum x_j = n\bar{x}$ but are otherwise independent. Taking as null hypothesis that the pattern is random, the probability of obtaining any value of the index of dispersion may then be found by consulting a table of percentage points of the χ^2-distribution. Unless the observed value of the index were improbably high, we should accept the null hypothesis.

However, it is often *un*reasonable to postulate that a pattern is random; or, if not positively unreasonable, there may at least be no particular grounds for favoring the hypothesis of randomness over any other imaginable pattern. The V/m ratio should then be regarded not as a test criterion but merely as a sample statistic descriptive of a population's pattern. It is important to distinguish between the use of V/m (or, equivalently, of the index of dispersion) as a test criterion and its use as a measure of aggregation when no hypothesis is being entertained. In the latter case V/m is simply an estimate of a population parameter on a par with an estimate of, say, mean density. If V/m is found to be close to 1, this should not prompt the conclusion that the pattern is truly random in the sense that the individuals are independent and the expected number per unit is the same for all units. The latter conclusion is justified only if there are a priori reasons for supposing it to be true and if application of the test gives no reason for rejecting it.

David and Moore (1954) have suggested, as a measure of aggregation, $I = (V/m) - 1$. They call I the "index of clumping" and describe a method for comparing two values of I, I_1 and I_2, say, from two different populations. The comparison may be made regardless of whether the means differ. Suppose samples of the same size n have been collected from both populations. Let m_1 and m_2 be the means of the two sets of observations and let V_1 and V_2 be their variances; then $I_j = (V_j/m_j) - 1$ with $j = 1, 2$. Now evaluate

$$w = -\tfrac{1}{2}\ln\left(\frac{V_1/m_1}{V_2/m_2}\right).$$

David and Moore state that if w lies outside the range of $-2.5/\sqrt{(n-1)}$ and $+2.5/\sqrt{(n-1)}$ then I_1 and I_2 differ significantly at the 5% level. Therefore, if we choose to regard I as a measure of aggregation, a method of comparing the degrees of aggregation of two populations is provided; but, as already explained, we may prefer to define aggregation in some other manner. So it would be truer to say that David and Moore's test is to judge the significance of the difference between two values of a certain function, I, of the means and variances. If and only if we have chosen this particular function as our preferred measure of aggregation can we interpret a difference between I_1 and I_2 as being equivalent to a difference between the two populations in their aggregation.

In examining the suitability of I as a measure of aggregation, it is interesting to visualize what would happen if, in a given population, a proportion $(1 - \theta)$ of the individuals were selected at random and killed or removed. Now consider the pattern of the survivors. Should this pattern be thought of as having the same or a lesser degree of aggregation when compared with that of the original population? Either answer is reasonable.

On the grounds that the greatest number of deaths will occur in what were originally the most densely populated units, with the result that the clumps are less dense than they were before, we could argue that the deaths had reduced aggregation. On the other hand, the facts that the survivors are still at their original locations and that the only change in the population has been the removal of randomly chosen individuals lead to the argument that a measure of aggregation should be used that is *not* affected by random deaths. Random deaths could be said to alter only the mean density of the population while leaving other aspects of its pattern unchanged. This shows that we are free to choose how aggregation is to be defined and that the properties of any particular measure depend on the definition.

We describe below a measure that remains unaltered when deaths take place at random among population members. First we shall show that David and Moore's I decreases linearly with decreasing population density (provided the deaths are at random) and that this is true whatever the initial frequency distribution.

Suppose the pgf of the initial distribution is $G_0(z)$ and of the final distribution, after the deaths, $G_1(z)$. Since θ is the probability that an individual will survive and is the same for all individuals, we have

$$G_1(z) = G_0(\theta z + 1 - \theta).$$

We now wish to find the first and second moments of $G_0(z)$ and $G_1(z)$ and determine how they are related. This is most easily done by first deriving the factorial moments. The factorial moment generating function (fmgf), denoted by $\Phi(u)$, is easily seen to be the same as $G(1 + u)$; that is,

$$\Phi(u) = \sum_{i=0}^{\infty} \frac{u^i}{i!} \mu'_{(i)} = G(1 + u),$$

where $\mu'_{(i)}$ is the ith factorial moment about the origin.

So, writing $\Phi_0(u)$ and $\Phi_1(u)$ for the fmgf of the initial and final distributions respectively, we have

$$\Phi_0(u) = G_0(1 + u)$$

and

$$\Phi_1(u) = G_1(1 + u) = G_0[\theta(1 + u) + 1 - \theta] = G_0(1 + \theta u) = \Phi_0(\theta u).$$

Now put $\mu'_{(i),\,0}$ and $\mu'_{(i),\,1}$ for the ith factorial moments of the initial and final distributions. Then, since $\Phi_1(u) = \Phi_0(\theta u)$,

$$\sum_0^{\infty} \mu'_{(i),\,1} \frac{u^i}{i!} = \sum_0^{\infty} \mu'_{(i),\,0} \frac{(\theta u)^i}{i!}.$$

Equating coefficients of $u^i/i!$ for $i = 1, 2$, we see that

$$\mu'_{(1),\,1} = \theta \mu'_{(1),\,0} \quad \text{and} \quad \mu'_{(2),\,1} = \theta^2 \mu'_{(2),\,0}.$$

Now, the first factorial moment of a distribution is identical with the mean m; and the variance V is given by

$$V = \mu'_{(2)} + \mu'_{(1)} - \mu'^2_{(1)} \quad \text{or} \quad \mu'_{(2)} + m - m^2.$$

So

$$I = \left(\frac{V}{m}\right) - 1 = \left(\frac{\mu'_{(2)}}{m}\right) - m,$$

and for the initial and final populations the indices of clumping are given by

$$I_0 = \left(\frac{\mu'_{(2),\,0}}{m_0}\right) - m_0 \quad \text{and} \quad I_1 = \left(\frac{\theta^2 \mu'_{(2),\,0}}{\theta m_0}\right) - \theta m_0 = \theta I_0.$$

Thus it follows that as the density of a population decreases owing to random deaths the index of clumping also decreases and is θ times its original value when a proportion θ of the original population remains.

This provides a method for determining whether the deaths in a cohort of sedentary organisms are density dependent, provided we can be certain there is no migration from unit to unit. Suppose the fate of an even-aged population of organisms (i.e., a cohort) were being followed; for instance, they might be pest caterpillars on tree shoots. The size of the population will gradually dwindle as time advances owing to the deaths of some of its members. If, at a succession of times, a sample is taken from the population of habitable units, we may determine how I varies with m. If there is no density dependence, the relation will be a straight line passing through the origin. If individuals belonging to dense aggregates are more likely to die than those in less crowded units, then I will decrease more rapidly; conversely, if there is inverse density dependence, that is, if living in dense clumps favors survival, I will decrease less rapidly.

3. The Negative Binomial Parameter, k

When the number of organisms per unit has a negative binomial distribution, we may use the parameter k of the series as a measure of aggregation (Waters, 1959). Since, for the negative binomial, $V = m + m^2/k$ (see page 85), in terms of David and Moore's index I, $k = m/I$; that is, low values of k indicate pronounced clumping and high values, slight clumping. To obtain an index of aggregation that increases with increasing clumping some authors use a function of k such as its reciprocal (Taylor, 1961).

An interesting property of k is that it remains unaltered when a population decreases in size owing to random deaths. This is the property we suggested earlier might be desirable in measures of aggregation: the measure may be thought of as representing some intrinsic property of a spatial pattern whatever the density:

To see this, consider a negative binomial series with pgf $(Q - Pz)^{-k}$ having mean $m = kP$ and variance $V = m(1 + m/k)$. Then

$$p_r = \binom{k+r-1}{r} \frac{P^r}{(1+P)^{k+r}},$$

where, for convenience, we have written $\binom{k+r-1}{r}$ in place of $[\Gamma(k+r)]/r!\Gamma(k)$. Now let a proportion $1 - \theta$ of the individuals be selected at random and destroyed so that a proportion θ survives in the final population.

Writing p'_r for the probability that a unit will contain r individuals in the final population, we tabulate initial and final probabilities (see Table 5). Clearly

$$p'_r = \theta^r \sum_{i=r}^{\infty} \binom{i}{r} (1-\theta)^{i-r} p_i .$$

TABLE 5

INDIVIDUALS PER UNIT i	INITIAL PROBABILITY p_i	FINAL PROBABILITY p'_i
0	p_0	$p'_0 = p_0 + (1-\theta)p_1 + (1-\theta)^2 p_2 + \cdots$
1	p_1	$p'_1 = \theta p_1 + \binom{2}{1}\theta(1-\theta)p_2 + \binom{3}{1}\theta(1-\theta)^2 p_3 + \cdots$
2	p_2	$p'_2 = \theta^2 p_2 + \binom{3}{2}\theta^2(1-\theta)p_3 + \binom{4}{2}\theta^2(1-\theta)^2 p_4 + \cdots$
$\cdots$	$\cdots$	$\cdots$
r	p_r	$p'_r = \theta^r p_r + \binom{r+1}{r}\theta^r(1-\theta)p_{r+1} + \binom{r+2}{r}\theta^r(1-\theta)^2 p_{r+2} + \cdots.$

Substituting the negative binomial term for p_i

$$
\begin{aligned}
p'_r &= \theta^r \sum_{i=r}^{\infty} \binom{i}{r}(1-\theta)^{i-r}\binom{k+i-1}{i}\frac{P^i}{(1+P)^{k+i}} \\
&= \theta^r\left\{\binom{k+r-1}{r}\frac{P^r}{(1+P)^{k+r}} + \binom{r+1}{r}(1-\theta)\binom{k+r}{r+1}\frac{P^{r+1}}{(1+P)^{k+r+1}} + \cdots\right\} \\
&= \frac{(\theta P)^r}{(1+P)^{k+r}}\binom{k+r-1}{r}\left\{1 + \binom{k+r}{1}\left(\frac{P(1-\theta)}{1+P}\right)\right. \\
&\qquad\left. + \binom{k+r+1}{2}\left(\frac{P(1-\theta)}{1+P}\right)^2 + \cdots\right\} \\
&= \binom{k+r-1}{r}(\theta P)^r\frac{1}{(1+P)^{k+r}}\left[1 - \frac{P(1-\theta)}{1+P}\right]^{-(k+r)} \\
&= \binom{k+r-1}{r}\frac{(\theta P)^r}{(1+\theta P)^{k+r}},
\end{aligned}
$$

and this is the rth term of a negative binomial series with parameters θP and k. The original mean was $m_0 = kP$, the new mean is $m_1 = \theta kP$, and the value of the exponent k is unchanged.

This is the reason for choosing k, or some function of it such as k^{-1}, as a measure of aggregation in patterns that yield a negative binomial distribution when sampled. However, this argument holds only for negative binomial populations. It is true that if we were to use as an (inverse) measure of aggregation the function k' defined as $k' = m^2/(V - m) = m/I$, then k' would remain unchanged if deaths occurred at random in a population. This is true whatever the parent distribution and follows from the fact that $k_1' = m_1/I_1 = \theta m_0/\theta I_0 = k_0'$, where, as before, the subscripts 0 and 1 denote initial and final values. Further, if the parent distribution were negative binomial, k' would be the moment estimator of the parameter k. However, when the parent distribution is not negative binomial, it is meaningless to speak of it as having the parameter k; and if we were to assume that a distribution was negative binomial when it was not and estimate "k" by maximum likelihood, this estimate $\hat{k}$ would *not* remain unchanged when random deaths occurred.

A numerical example illustrates this point. The initial frequencies shown in Table 6 have been chosen arbitrarily and do not represent any theoretical

TABLE 6

	INITIAL FREQUENCY	FINAL FREQUENCY
i	np_i	np_i'
0	40	63.59
1	30	44.06
2	10	40.78
3	20	31.88
4	30	15.16
5	40	4.06
6	30	0.47
	$n = 200$	200.00

contagious distribution. The final frequencies were obtained by assuming that half the population had been removed at random. In Table 6:

$$m_0 = 3.05, \qquad m_1 = 1.525,$$
$$V_0 = 4.7475, \qquad V_1 = 1.9494,$$

$$I_0 = \frac{V_0}{m_0} - 1 = 0.557, \qquad I_1 = \frac{V_1}{m_1} - 1 = 0.278\ (= \tfrac{1}{2} I_0),$$

$$k_0' = k_1' = 5.480,$$
$$\hat{k}_0 < 3.0, \qquad \hat{k}_1 > 4.0.$$

The maximum likelihood estimates $\hat{k}_0$ and $\hat{k}_1$ were calculated by the method described by Bliss and Fisher (1953), and it is seen that $\hat{k}_1 > \hat{k}_0$. Thus, if we had assumed, mistakenly, that the parent distribution was negative binomial, we should be led to conclude that aggregation had decreased as a result of the deaths; this, in turn, would suggest that the deaths had been density dependent.

4. Lloyd's Indices of Mean Crowding and Patchiness

On page 92 we advanced arguments in favor of using a measure of aggregation that (a) does and (b) does not change when some of the population members are removed at random. Obviously, two different things are being envisaged and each should be measured separately. Lloyd (1967) has proposed an "index of mean crowding" and an "index of patchiness" that seem to meet the requirements. He defines mean crowding as the mean number per individual of other individuals in a unit; these other individuals may be thought of as co-occupants of the unit with the first individual. The "crowding" $\overset{*}{m}$ is an average over individuals instead of over units. It is calculated by counting, for each individual in a total population of N individuals, say, the number of co-occupants X_i that share the unit with it; $(i = 1, 2, \ldots, N)$.

Then the mean crowding is

$$\overset{*}{m} = \frac{1}{N} \sum_{i=1}^{N} X_i .$$

If there is a total of n units and $x_j (j = 1, 2, \ldots, n)$ denotes the number of individuals in the jth unit, then

$$\sum_{i=1}^{N} X_i = \sum_{j=1}^{n} x_j(x_j - 1); \qquad \text{also} \sum_{j=1}^{n} x_j = N.$$

Then

$$\overset{*}{m} = \frac{\sum x_j^2}{\sum x_j} - 1.$$

In terms of the mean and variance of the x's, m and V, say,

$$\overset{*}{m} = m + \left(\frac{V}{m} - 1\right) \quad \text{or} \quad m + I,$$

since

$$\frac{\sum x_j^2}{\sum x_j} = \frac{V + m^2}{m}.$$

Thus the crowding is numerically equal to the sum of the mean density and David and Moore's index of clumping I. As with I itself, $\overset{*}{m}$ must remain proportional to density as a population is depleted by random deaths. Using, as before, the subscripts 0 and 1 for the initial and final values in a dwindling population, we have

$$\overset{*}{m}_0 = m_0 + I_0, \qquad \overset{*}{m}_1 = m_1 + I_1.$$

When only a proportion θ of the initial population survives, we know that

$$m_1 = \theta m_0 \quad \text{and} \quad I_1 = \theta I_0, \qquad \text{whence } \overset{*}{m}_1 = \theta \overset{*}{m}_0.$$

The patchiness is defined as $\overset{*}{m}/m$, or the ratio of mean crowding to mean density. Random deaths leave patchiness unaltered, since $\overset{*}{m}_0/m_0 = \overset{*}{m}_1/m_1$; this remains true whatever the form of the parent distribution. Notice also that

$$\frac{\overset{*}{m}}{m} = 1 + \frac{I}{m} = 1 + \frac{1}{k'}.$$

Thus crowding is something that is experienced by each individual and depends on the total number present. Patchiness, on the other hand, is a property of a spatial pattern considered by itself without regard to density, and two populations can exhibit the same degree of patchiness even though their densities differ.

In the numerical example already given (page 96) it can be seen that the crowding decreases from $\overset{*}{m}_0 = 3.607$ to $\overset{*}{m}_1 = 1.803$, when half the population is destroyed. The patchiness remains constant and equal to 1.182.

9

The Pattern of Individuals in a Continuum

1. Introduction

So far we have considered only the spatial patterns of organisms that occupy small isolated units; the space available to the organisms was discrete. Now we turn to the case in which an extended continuum, either an area or a volume, is available to the organisms, and they may be found anywhere throughout it; this is Case 2 on page 81. Examples are individual plants scattered over an area of ground or microarthropods dispersed through a volume of soil. There are now no natural sampling units such as the discrete habitable units afforded and sampling units have to be arbitrarily defined. To fix ideas we suppose that the population to be studied consists of all the plants growing on an apparently homogeneous tract of level ground. It is assumed, further, that the plants are small in relation to the space available to them, that they are roughly equal in size, and that they reproduce entirely by seed and never vegetatively so that there is no difficulty in recognizing true individuals.

The commonest method of investigating the pattern of such a population is to sample it with randomly placed quadrats, which are small sample areas that are usually, but not necessarily, square. We then count the number of individual plants in each quadrat, compile a frequency table to show the observed numbers of quadrats that contained 0, 1, 2, . . . , individuals, and examine the observed distribution. The method has two great drawbacks. In the first place, the results are greatly affected by the size of quadrat used; and, in the second, when the observations from all the quadrats are pooled to compile a frequency table, no record is kept of the locations of the quadrats. Even though *spatial* pattern is ostensibly being studied, the spatial relationship of the sparsely occupied quadrats and densely occupied quadrats is often ignored. The records normally kept do not show, for instance, whether sparse and dense quadrats were randomly mingled with one another or whether, instead, they tended to occur as contiguous groups on the ground.

We now consider how the size of the quadrats affects the results of quadrat sampling.

2. The Effect of Quadrat Size

To say that plants on the ground have a random pattern or are randomly dispersed is equivalent to saying that every point on the ground (within the area of study) is as likely as every other to be the site of an individual plant. When such a pattern is sampled with randomly thrown quadrats, the expected distribution of the number of plants per quadrat is Poisson with parameter λ, where λ is the mean number of plants per quadrat. Changing the quadrat size simply alters the magnitude of λ, which is proportional to quadrat size, and the distribution remains Poisson for all quadrat sizes. Now suppose that the plants are clumped: on certain patches of ground their density is high and on others, low. When such a pattern is sampled, the results will, in general, be influenced by quadrat size. A large quadrat will often contain the whole of a densely occupied patch or even several of these patches, whereas a small quadrat may contain only part of a dense patch. As a result, measures of aggregation based on quadrat data will not as a rule be unique; different values will be obtained with different quadrat sizes.

There are, however, two rather special types of pattern for which this does not hold; that is, if the measure of aggregation to be used is suitably chosen, it will not vary with quadrat size.

1. Consider first a population that occurs in the form of clumps so compact and widely spaced that they are hardly ever cut through by the edges of a quadrat. If a measure of aggregation is used that depends only on the parameter (or parameters) of the distribution of the number of individuals per clump, the measure will be unaffected by changes in quadrat size; for example, suppose that the number of clumps per quadrat is a Poisson variate with parameter λ_1, say, and the number of individual plants per clump is a Poisson variate with parameter λ_2. Then the distribution of the number of plants per quadrat will be Poisson-Poisson with parameters λ_1 and λ_2 (see page 83). The mean and variance of this distribution are $\lambda_1\lambda_2$ and $\lambda_1\lambda_2(1 + \lambda_2)$, respectively, and the V/m ratio is thus $1 + \lambda_2$. Equivalently, David and Moore's index of clumping, I, has the value λ_2. In these circumstances, then, I is identical to the mean number of plants per clump. Obviously, changes in quadrat size will affect only λ_1 and leave λ_2 unaltered, and we see that by using I (or the V/m ratio) we shall have a measure of aggregation that does not depend on quadrat size. This argument applies, of course, only if we assume that the frequency with which clumps are cut through by quadrat edges is negligibly small. However, if the clumps were compact enough for

this assumption to be justified, they would presumably be sufficiently distinct to be easily recognizable. We could then investigate the spatial pattern of the clumps and, separately, the distribution of the number of plants per clump. It would be futile to treat the number of plants per quadrat as a variate to be investigated and needless to attempt indirect inferences based on such observations.

2. Next, consider a pattern consisting of a mosaic, or patchwork, of several phases. Assume that the pattern is very coarse; that is, the patches are large in relation to the size of the quadrats. Within any one phase, which may be represented by a number of distinct patches, the pattern is random, but different phases have different densities or Poisson parameters. Now let the area be sampled with quadrats small enough for us to assume that nearly all of them lie wholly within one or another of the patches and not across patch boundaries. The resultant distribution of plants per quadrat will obviously consist of a mixture of Poisson distributions. Suppose that there are k different phases and that in the jth phase, which occupies a proportion π_j of the total area, the Poisson parameter is $\lambda_j (j = 1, 2, \ldots, k)$. The probability that a quadrat will contain n plants is then

$$P_n = \frac{\pi_1 \lambda_1{}^n e^{-\lambda_1}}{n!} + \frac{\pi_2 \lambda_2{}^n e^{-\lambda_2}}{n!} + \cdots + \frac{\pi_k \lambda_k{}^n e^{-\lambda_k}}{n!}$$

or

$$P_n = \pi_1 p_{1n} + \pi_2 p_{2n} + \cdots + \pi_k p_{kn}, \tag{9.1}$$

where we have put $\lambda_j{}^n e^{-\lambda_j}/n! = p_{jn}$ for brevity. Evidently, if the quadrat size is changed (but still kept so small that there is a negligible chance that any quadrat will lie across patch boundaries), the only change in the resultant distribution will be that stemming from the fact that each of the Poisson parameters will be multiplied by a constant factor. Therefore it should be possible to find some index of aggregation that is unaffected by this change. Lloyd's (1967) index of patchiness meets this requirement, as we now show.

Consider the distribution whose general term is given by (9.1). Its mean and variance may be found as follows: the mean is

$$m = \sum_{j=1}^{k} \left(\pi_j \sum_{n=1}^{\infty} n p_{jn} \right) = \sum_{j=1}^{k} \pi_j \lambda_j .$$

Similarly, the second moment about the origin is

$$\sum_{j=1}^{k} \left(\pi_j \sum_{n=1}^{\infty} n^2 p_{jn} \right) = \sum_{j=1}^{k} \pi_j (\lambda_j + \lambda_j{}^2),$$

whence the variance V is

$$V = \sum_{j=1}^{k} \pi_j (\lambda_j + \lambda_j{}^2) - \left(\sum_{j=1}^{k} \pi_j \lambda_j \right)^2 .$$

Recalling that the patchiness (see page 98), which we here denote by C, is

$$C = 1 + \frac{V - m}{m^2},$$

we now have

$$C = 1 + \frac{\sum \pi_j \lambda_j^2 - (\sum \pi_j \lambda_j)^2}{(\sum \pi_j \lambda_j)^2} = \frac{\sum \pi_j \lambda_j^2}{(\sum \pi_j \lambda_j)^2}.$$

If the population is sampled again with quadrats r times the size of those originally used, the Poisson parameter in the jth phase will become $r\lambda_j$ for all j. Denoting by C_r the patchiness that is observed when quadrats of r units of area are used, we see that

$$C_r = \frac{\sum \pi_j (r\lambda_j)^2}{(\sum \pi_j r\lambda_j)^2} = C,$$

and thus the patchiness is the same as before.

Had we used the index of clumping I instead of the patchiness as a measure of aggregation, it would have been multiplied by the factor r. Writing I and I_r for the index of clumping obtained with quadrats of 1 and r units of area, respectively, it is seen that

$$I = \frac{V - m}{m} = \frac{\sum \pi_j \lambda_j^2 - (\sum \pi_j \lambda_j)^2}{\sum \pi_j \lambda_j}.$$

If we multiply every λ_j by the factor r, the index becomes

$$I_r = \frac{\sum \pi_j (r\lambda_j)^2 - (\sum \pi_j r\lambda_j)^2}{\sum \pi_j r\lambda_j} = rI.$$

Similarly, Lloyd's index of crowding, $\overset{*}{m}$, is multiplied by the factor r. With quadrats of unit area the crowding is $\overset{*}{m} = m + I$, and with quadrats of r units of area it is $\overset{*}{m}_r = r(m + I) = r\overset{*}{m}$.

It is interesting to note that Lloyd's "patchiness" is almost identical with Morisita's (1959) "index of dispersion," I_δ. Beginning from the same premise that the population under investigation consists of a mosaic of large (relative to the quadrats) patches within which the pattern is random, Morisita also sought for an index that would be unaffected by changes in quadrat size. He derived his index I_δ from entirely different considerations, however. To explain his derivation it is necessary first to mention the notion of "diversity" and Simpson's (1949) method of measuring it.

Suppose we have a collection of N objects of s different kinds, of which n_1 are of the first kind, n_2 of the second kind, ... and n_s of the sth kind, with

$\sum n_i = N$. If two objects are picked at random, and without replacement, from the whole collection, the probability that both will be of the same kind is clearly

$$\frac{\sum_{i=1}^{s} n_i(n_i - 1)}{N(N-1)}.$$

It is reasonable to call the diversity of the collection great if this probability is low and slight if it is high. Now suppose we sample a population with s quadrats and attach a label to each of the N individuals encountered to show in which of the quadrats it was found. It is assumed that there is no overlapping of the quadrats. Let x_i of the individuals be from the ith quadrat ($i = 1, \ldots, s, \sum_i x_i = N$); then these x_i individuals are classified as being of the ith kind, since they belong to the ith quadrat. The probability that any two individuals chosen at random (from the total of N individuals) belong to the same quadrat is therefore

$$\delta = \frac{\sum_i x_i(x_i - 1)}{N(N-1)}.$$

If the individuals are crowded into comparatively few of the quadrats, that is, if they are aggregated, δ will be high; conversely, if the individuals are fairly uniformly spaced so that they are more or less evenly apportioned among the s quadrats, δ will be low.

Consider now the expected value of δ when a population of random pattern is sampled. In this case the probability that a randomly selected individual will come from a given quadrat is the same for all quadrats and is therefore $1/s$. Since the individuals are independent of one another, the same is true for a second individual. Therefore the probability that two randomly picked individuals will both come from the same given quadrat is $1/s^2$. Summing over all quadrats, we see that the expected value of δ, given a random pattern, is $\delta_{\text{ran}} = \sum (1/s^2) = 1/s$. Morisita's I_δ is defined as $\delta/\delta_{\text{ran}} = s\delta$ and so has the value 1 in a random pattern. In an aggregated pattern, in which a high proportion of the individuals is concentrated into only a few of the quadrats, $I_\delta > 1$.

It may be seen that

$$I_\delta = \frac{s}{N-1} \frac{\sum_{i=1}^{s} n_i(n_i - 1)}{N} = \frac{s}{N-1} \overset{*}{m} = \frac{N}{N-1} \cdot C,$$

since $C = \overset{*}{m}/m$ and $sm = N$.

It follows that if we know that a pattern consists of a mosaic of patches with different densities, within each of which the individuals are randomly dispersed, either C or I_δ may be used as a measure of aggregation and that, provided the quadrats are sufficiently small, the values will be independent of quadrat size. To assume without evidence, however, that a pattern is of mosaic form is usually unjustified, but we can test the validity of the assumption by sampling the population repeatedly, using quadrats of several sizes and judging whether C (or I_δ) remains constant for several of the smallest quadrats.

Much can indeed be learned about a pattern by examining the way in which some measure of aggregation varies with quadrat size. Besides showing that for mosaic patterns with large patches the I_δ versus quadrat size curve is horizontal when the quadrats are small, Morisita (1959) also discusses the form this curve will take, given other types of mosaic pattern. Its shape depends on the sizes of the patches and on the pattern of the individuals within the patches.

We shall not explore the matter further here but shall turn to a consideration of Greig-Smith's method of pattern analysis.

3. Grids of Contiguous Quadrats

Greig-Smith's (1952, 1964) method of pattern analysis is also based on the fact that the way in which a measure of aggregation varies with quadrat size provides information on pattern.

To sample an area over and over again with quadrats of successively larger sizes is exceedingly time-consuming, not to mention the fact that the vegetation becomes increasingly trampled as work proceeds. Greig-Smith therefore proposes the use of a grid, or lattice, of small square unit cells that completely cover the area to be studied. The grid units are, in fact, contiguous quadrats. The number of individuals in each grid unit is first counted. Adjacent pairs of units are then combined to give oblong two-unit blocks which are twice as big and half as numerous as the original units. Adjacent two-unit blocks are next combined to give square four-unit blocks; now these are combined to give oblong eight-unit blocks, and so on. In this way from a single examination of the area a sequence of "quadrat" sizes is obtained in which the quadrat area is doubled at each step.

The total sum of squares about the mean for the single grid cells (or one-unit blocks) may now be apportioned as in an analysis of variance. We obtain sums of squares that derive from the difference between each pair of single units within the two-unit blocks, from the difference between each pair of two-unit blocks within the four-unit blocks, ... from the difference between each pair of 2^j-unit blocks within the 2^{j+1}-unit blocks, and so on. These within-blocks sums of squares may be reduced to mean squares that are equivalent to V/m ratios. Evidently, while the blocks are still very small in relation to the mosaic patches of the pattern, adjacent half-blocks within each block are both likely to be inside a single patch and the mean square will be low. Thereafter, as block size increases, so will the mean square until, according to Greig-Smith, a block size is reached equal to the area of the patches. If block size is increased still further, the mean square will remain at this high level if the patches themselves are random or aggregated; but if the patches have a regular arrangement, the mean square will fall off again. Hierarchical clumping will produce a succession of peaks in the mean-square-versus-block-size graph.

Although this method of studying pattern has been favored by a number of plant ecologists [e.g., Phillips (1953), Kershaw (1960), Cooper (1961)], it has several drawbacks and it is worthwhile to list them:

1. Because each block is formed by successive combinations of the original grid units, the calculated mean squares are not independent. So we cannot perform variance ratio tests to determine whether a large mean square exceeds a small one significantly. The mean-square-versus-block-size graphs can thus be judged only subjectively.

2. The method can be used only on areas small enough to be examined in their entirety. The whole of the area to be investigated must be included in the grid.

3. The graph sometimes has a sawtoothed shape because oblong blocks consistently give mean squares less than those of the square blocks on either side of them in the sequence of sizes.

4. Block size is doubled at each step; therefore there is no means of knowing what the form of the graph would have been had blocks of intermediate sizes been used as well. A peak for blocks of 16 units, say, can be interpreted as meaning only that the mean patch area lies somewhere between 8 and 32 units. Also, it is not clear what sort of graph should be expected if the patches vary greatly in size.

5. Consider a population formed of a number of patches (or clumps) of individuals dispersed over otherwise empty ground and visualize also the same pattern in reverse: what were clumps in the first population are now lacunae and what was empty ground is now occupied by randomly dispersed individuals with the same density as those that formerly occupied the patches (see Figure 12). What might be called the "grain" of these two patterns is the same, and, as a result, they have very similar mean-square-versus-block-size graphs when analyzed by Greig-Smith's method. The maps of these artificial populations shown in the figure were analysed and the result is also shown; it can be seen that for blocks of less than 128 units the graphs are alike, although in map B clumps in the ordinary sense do not exist. The divergent results at the largest block sizes occur simply because the difference in numbers of individuals between the left and right halves of the areas is greater in map B than in map A.

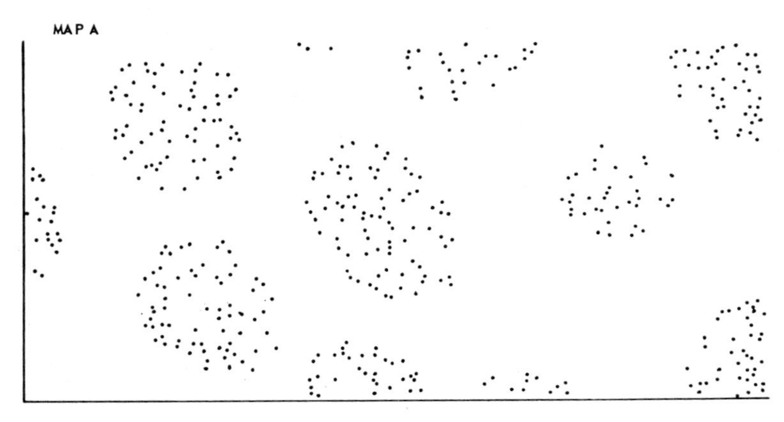

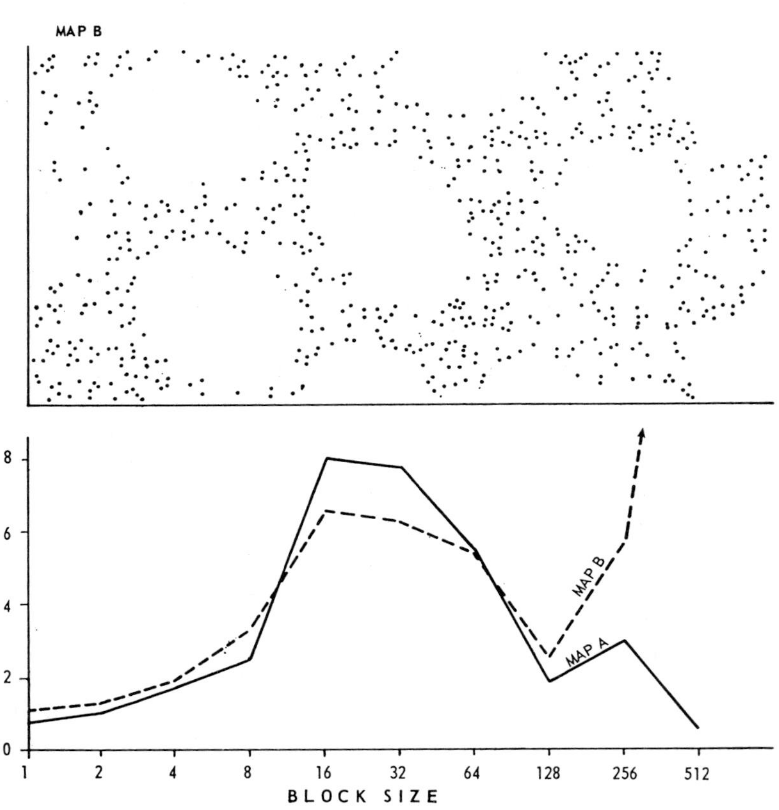

Figure 12. Two patterns (artificial) that give very similar graphs of mean square versus block size.

4. Random Mingling of Sparse and Dense Grid Cells

Consider again a grid map of a population's pattern. As before, the map shows the number of individuals in each cell of the grid. Let us now subdivide the cells into two classes, say, sparse and dense, and color them white and black, respectively. If many cells are empty, the subdivision may be into empty versus occupied cells; or if nearly all the cells are occupied, the subdivision may be into cells containing fewer than x individuals (sparse cells) and cells containing x or more individuals (dense cells), with x being assigned a value that ensures a fairly equal division of the cells into whites and blacks.

It is clear that, regardless of the frequency distribution of the number of individuals per cell, indeed even if this distribution were well fitted by a Poisson series, we cannot regard the pattern as random unless the black and white cells are randomly mingled. Therefore it is worthwhile to consider how we may test for random mingling. A way of doing this has been proposed by Krishna Iyer (1949). To exemplify the use of his test he considered the mingling of diseased and healthy plants in a regular plantation in which all the plants were at lattice points. The method is, however, equally good for determining whether the dense (black) and sparse (white) cells in a grid map are randomly mingled.

Suppose the grid is rectangular, with m rows and n columns. There are then, $mn = b$ say, cells; we also write $m + n = a$. Assume that r_1 of the cells are black and r_2 are white so that $r_1 + r_2 = b$. The test consists in comparing with expectation the observed number of black-black joins: a black-black join occurs when two black cells adjoin each other within a row or column or diagonally. Thus in the example shown in Figure 13 $m = 5$, $n = 6$ and therefore $a = 11$ and $b = 30$. The number of black cells is $r_1 = 15$ and the number of white cells is $r_2 = 15$. There are 19 black-black joins, as shown by the arrows. To determine whether the observed number of black-black joins exceeds expectation significantly we must first find the mean and variance of the distribution of the number of black-black joins on the null hypothesis of random mingling of the black and white cells.

We begin by noting that the kth factorial moment of the distribution is $k!$ times the sum of the probabilities of the different ways of obtaining k black-black joins in the grid. [For a proof of this see Krishna Iyer (1952).]

The probability that any two given cells will be black is obviously $r_1(r_1 - 1)/[b(b - 1)]$. Thus by putting A for the number of ways of choosing two cells so that they are adjoining, the first factorial moment of the number of black-black joins is

$$\mu'_{(1)} = \frac{1!Ar_1^{(2)}}{b^{(2)}},$$

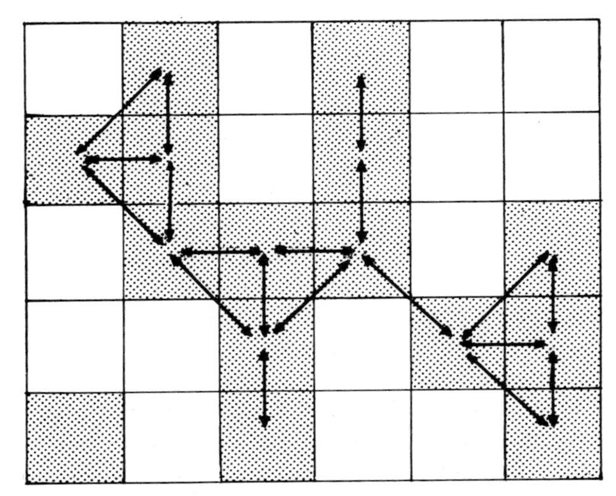

Figure 13. A grid map of an area in which the grid cells have been classified as dense (stippled) and sparse (white). The arrows show the joins between adjoining dense cells.

where $r_1^{(x)} = r_1(r_1 - 1) \cdots (r_1 - x + 1)$ and similarly for $b^{(x)}$. This follows since for each of the A possible ways of choosing a pair of adjoining cells the probability is the same, namely $r_1^{(2)}/b^{(2)}$, that both members of the pair will be black.

Likewise, two black-black joins will occur (a) whenever three adjoining cells (not necessarily in a straight line) are all black and (b) whenever there are four black cells in the form of two adjoining pairs. (The fact that in these cases there may be more than two joins is irrelevant.) For case (a) the probability that any three given cells will be black is $r_1^{(3)}/b^{(3)}$ and we denote by B the number of ways in which these three cells may be chosen so that they will be connected by at least two joins. For case (b) the probability that any four given cells will be black is $r_1^{(4)}/b^{(4)}$ and we denote by C the number of ways of choosing two pairs of adjoining cells. Then the second factorial moment about the origin of the distribution is

$$\mu'_{(2)} = 2!\left[B\frac{r_1^{(3)}}{b^{(3)}} + C\frac{r_1^{(4)}}{b^{(4)}}\right].$$

It is now necessary to determine A, B, and C. Consider again the grid of $mn = b$ cells. It has four corner cells, $2(m + n - 4)$ cells that are on the edges of the grid but not at the corners, and $(m - 2)(n - 2)$ interior cells.

To determine A note that if any one corner cell is black there are three ways of choosing an adjoining cell that could also be black. If any one edge cell is black, there are five adjoining cells that could also be black, and if any interior cell is black there are eight adjoining cells that could also be black. Therefore $2A = 4 \times 3 + 2(m + n - 4) \times 5 + (m - 2)(n - 2) \times 8$; (the factor 2 on the left-hand side occurs because all joins have been counted twice). Then

$$A = 2 - 3(m + n) + 4mn$$

or

$$A = 2 - 3a + 4b.$$

Next we find B. If a corner cell is to be connected by joins to two other cells, there are $\binom{3}{2}$ ways of choosing the two others; if an edge cell is to be joined to two others, there are $\binom{5}{2}$ ways of choosing the two adjoining cells; and for an interior cell there are $\binom{8}{2}$ ways of choosing the two adjoining cells.
Then

$$\begin{aligned} B &= 4\binom{3}{2} + 2(m + n - 4)\binom{5}{2} + (m - 2)(n - 2)\binom{8}{2} \\ &= 44 - 36a + 28b. \end{aligned}$$

To find C we note that there are A ways of obtaining a single adjoining pair of cells, hence $\binom{A}{2}$ ways of obtaining two such pairs; but these include the trios already discussed, of which there are B. Thus there are $\binom{A}{2} - B$ ways of obtaining a couple of distinct pairs (i.e., four cells altogether).
Therefore

$$C = \binom{A}{2} - B.$$

It now follows that the mean and variance of the number of black-black joins, on the null hypothesis that the black and white cells are randomly mingled, are

$$m = \mu'_{(1)} = \frac{A r_1^{(2)}}{b^{(2)}}$$

and

$$V = 2B \frac{r_1^{(3)}}{b^{(3)}} + 2C \frac{r_1^{(4)}}{b^{(4)}} + m - m^2,$$

since

$$V = \mu'_{(2)} + m - m^2.$$

The distribution is asymptotically normal. Although it seems to be unknown how large m and n must be for an assumption of normality to give a good approximation, Krishna Iyer uses the test for a lattice with $m = 15$ and $n = 20$. If, then, we have a grid with a sufficient number of cells, it is easy to test whether the black and white cells (or dense and sparse quadrats) are randomly mingled.

If the cells were very small and the pattern aggregated, we should expect an excessive number of black-black joins, since any two adjoining cells are likely to be both within or both outside of the densely populated patches. With large cells, on the other hand, even if the pattern were nonrandom the grid might be so coarse that the number of individuals in any one cell would be independent of the number in the cells adjoining it. The sparse and dense cells would then be randomly mingled.

It is not permissible (owing to dependence among the results) to sample the same area repeatedly with coarser and coarser grids in search of a cell size at which evidence of aggregation would vanish, but, if we wished to test a previously held hypothesis about the grain of a tract of vegetation, the test could be useful. It has not been used in this way so far by ecologists.

10

Studying Pattern by Distance Sampling

1. Introduction

Quadrat sampling is only one of the methods that may be used to study the spatial pattern of a population of organisms dispersed over a continuous surface. As we have remarked, it suffers from the grave disadvantage that quadrats are not natural sampling units but are necessarily arbitrary. There is, however, a wholly different method of investigating the pattern of points in a plane. This is by so-called "plotless sampling." What is examined is the spacing of the individuals, and there are two ways to proceed. We may locate sampling points at random throughout the area and measure the distance from each point to the individual nearest it, or, alternatively, select individuals at random from the whole population and from each of these measure the distance to its nearest neighboring individual. In either case the data consist of an empirical frequency distribution of a continuous variate, distance. Moreover, if the pattern of the population is random, the results are unaffected by whether random points or random individuals are used as the origins of measurement.

2. The Distribution of Distance-to-Neighbor in a Randomly Dispersed Population

Consider the distance from a random point (or a randomly chosen individual) to its nearest neighbor. We wish to find the probability distribution of this distance, given that the pattern of the population is random.

The required distribution function (cumulative distribution function) is, by definition,

$$F(r) = \Pr(\text{distance to nearest neighbor} \leq r).$$

This is the probability that a circle of radius r centered on the point will contain at least one individual or, equivalently, that this circle is not empty. Therefore

$$F(r) = 1 - e^{-\lambda r^2},$$

where the Poisson parameter λ denotes the mean number of individuals per circle of unit radius. The term $e^{-\lambda r^2}$ is, of course, the probability that a Poisson variate will take the value 0.

The probability density function (or frequency function) of the distribution and its mean and variance follow immediately. Thus the probability density function (pdf) is

$$f(r) = F'(r) = 2\lambda r e^{-\lambda r^2}.$$

The mean is

$$E(r) = \int_0^\infty 2\lambda r^2 \, e^{-\lambda r^2} \, dr.$$

Using the substitution $r^2 = x$, hence $dr = dx/2\sqrt{x}$ gives

$$E(r) = \lambda \int_0^\infty \sqrt{x} \, e^{-\lambda x} \, dx$$

$$= \frac{1}{2}\left(\frac{\pi}{\lambda}\right)^{1/2} \qquad (10.1)$$

The second moment about the origin is

$$E(r^2) = \int_0^\infty 2\lambda r^3 \, e^{-\lambda r^2} \, dr = \frac{1}{\lambda},$$

whence the variance is

$$\text{var}(r) = E(r^2) - [E(r)]^2 = \frac{4 - \pi}{4\lambda}. \qquad (10.2)$$

It is simpler and more convenient to use the square of the distance rather than the distance itself as variate. Putting $\omega = r^2$, the distribution function of the squared distance is then

$$F(\omega) = 1 - e^{-\lambda \omega}$$

and the pdf is

$$f(\omega) = \lambda e^{-\lambda \omega}.$$

The mean and variance of ω are easily found to be

$$E(\omega) = \frac{1}{\lambda} \quad \text{and} \quad \text{var}(\omega) = \frac{1}{\lambda^2}.$$

We next inquire into the sampling distribution of $\bar{\omega}$, the mean of a random sample of n values of ω. Since the distribution function of ω is

$$F(\omega) = 1 - e^{-\lambda\omega},$$

if we put $y = 2\lambda\omega$, the distribution function and the pdf of y are, respectively,

$$G(y) = 1 - e^{-y/2} \quad \text{and} \quad g(y) = \tfrac{1}{2}e^{-y/2}.$$

We now obtain the moment generating function (mgf) of y, namely,

$$\begin{aligned} E(e^{ty}) &= \tfrac{1}{2}\int_0^\infty e^{-y/2}e^{ty}\,dy \\ &= (1-2t)^{-1}. \end{aligned}$$

It will be seen that this is the mgf of the χ^2-distribution with two degrees of freedom (e.g., Hoel, 1954).

We can now obtain the mgf of $n\bar{y}$, the sum of n independent values of y, since it is given by

$$[E(e^{ty})]^n = (1-2t)^{-n},$$

and this is the mgf of the χ^2-distribution with $2n$ degrees of freedom. This, then, is the distribution possessed by $n\bar{y} = 2n\lambda\bar{\omega}$; that is, $2n\lambda\bar{\omega} = z$, say, has pdf

$$h(z) = \frac{1}{2^n\Gamma(n)}\,z^{n-1}e^{-z/2},$$

which is the pdf of the χ^2-distribution with $2n$ degrees of freedom. Then $\bar{\omega} = z/2n\lambda$ has pdf

$$\begin{aligned} f(\bar{\omega}) &= 2n\lambda\frac{1}{2^n\Gamma(n)}\,(2n\lambda\bar{\omega})^{n-1}e^{-n\lambda\bar{\omega}} \\ &= \frac{(n\lambda)^n\bar{\omega}^{n-1}e^{-n\lambda\bar{\omega}}}{\Gamma(n)}, \end{aligned} \tag{10.3}$$

which is the distribution we set out to determine.

Since $2n\lambda\bar{\omega}$ has a χ^2-distribution with $2n$ degrees of freedom and since, also, for the χ^2-distribution the expectation is equal to the number of degrees of freedom, it follows that

$$E(2n\lambda\bar{\omega}) = 2n \quad \text{or} \quad E(\bar{\omega}) = \frac{1}{\lambda}. \tag{10.4}$$

This suggests that we might use $1/\bar{\omega}$ as an estimator of λ, the mean number of individuals per circle of unit radius. However, it is a biased estimator, as Moore (1954) has shown.

Thus

$$E\left(\frac{1}{\bar{\omega}}\right) = \int_0^\infty \frac{1}{\bar{\omega}} \frac{(n\lambda)^n \bar{\omega}^{n-1} e^{-n\lambda\bar{\omega}}}{\Gamma(n)} \cdot d\bar{\omega}$$

$$= \frac{(n\lambda)^n}{\Gamma(n)} \int_0^\infty \bar{\omega}^{n-2} e^{-n\lambda\bar{\omega}}\, d\bar{\omega}$$

$$= \frac{n}{n-1} \cdot \lambda.$$

So to obtain an unbiased estimator, $\tilde{\lambda}$, say, from the mean of n observed values of ω we should need to put

$$\tilde{\lambda} = \frac{n-1}{n} \cdot \frac{1}{\bar{\omega}}. \tag{10.5}$$

Unfortunately, it is not possible in practice to estimate the density of a natural population of organisms by substituting an observed value of $\bar{\omega}$ in (10.5). The formula $(n-1)/n\bar{\omega}$ is an estimator of the density only when the population has a random pattern, so it can be used only if we can safely assume, or know beforehand, that a pattern is indeed random. However, we are never justified in assuming randomness without evidence for it; the assumption must always be tested. It turns out that there is no method of performing such a test by using only an observed sample of ω values. Even if an empirical distribution of ω values appears to be well fitted by the theoretical distribution $f(\omega) = \lambda e^{-\lambda\omega}$ it cannot be concluded that the pattern is random, since empirical distributions from nonrandom patterns are often not distinguishably different from this negative exponential form. Randomness therefore cannot be assumed without a test based on additional observations, and these observations entail estimating, or determining, the density; that is, distance measurements alone are not enough, and we must also carry out quadrat sampling or a complete count of the population. Thus an estimate of density has to be obtained *before* a test for randomness can be made, and then there is no longer any need to use the distance measurements for density estimation. The fact that it is impossible to base a density estimate on distance measurements and nothing else is a pity, since such a method would be valuable, especially to foresters who wish to estimate the density of the trees in a tract of forest.

However, distance measurements can be used in combination with density estimates in tests for randomness and in devising measures of nonrandomness that are independent of quadrat size.

3. Tests for Randomness Based on Distance Measurements

Three tests for randomness are described here and for conciseness we assume that the individuals are plants.

1. The test due to Hopkins and Skellam (1954) hinges on the fact that if, and only if, a pattern is random, the distribution of the distance from a random point to its nearest plant is identical with the distribution of the distance from a random plant to its nearest neighbor.

Denote by ω_1 the square of a point-to-plant distance and by ω_2 the square of a plant-to-neighbor distance and suppose a sample is obtained of n distances of each kind. The statistic $A = \sum \omega_1 / \sum \omega_2$ then has an expected value of 1 if the pattern is random and A may be used as a measure of nonrandomness. Clearly, if the plants are aggregated, we shall have $A > 1$; conversely, if they are more evenly spaced than in a randomly dispersed population, $A < 1$.

To determine whether A departs significantly from its expected value of 1 we determine the sampling distribution of

$$x = \frac{A}{1+A} = \frac{\sum \omega_1}{\sum \omega_1 + \sum \omega_2}.$$

Since $E(A) = 1$, $E(x) = \frac{1}{2}$.

To find the desired sampling distribution put

$$x = \frac{\sum \omega_1}{\sum \omega_1 + \sum \omega_2} = \frac{a}{a+b}, \text{ say.}$$

For convenience we may take $\lambda = 1$; this entails no loss of generality, since we are always at liberty to choose the unit of distance in such a way that the mean number of plants per circle of unit radius is 1. Then, using (10.3), it is seen that $\sum \omega_1 = n\bar{\omega}_1 = a$ has pdf

$$f(a) = \frac{a^{n-1}e^{-a}}{\Gamma(n)};$$

likewise $\sum \omega_2 = n\bar{\omega}_2 = b$ has a pdf of the same form. Now consider $a = bx/(1-x)$. The probability that, for fixed b, a will lie in the interval da, hence x in the interval dx, is

$$f(a)\,da = \frac{a^{n-1}e^{-a}}{\Gamma(n)}\,da,$$

which is equivalent to

$$g(x)\,dx = \frac{1}{\Gamma(n)}\left(\frac{bx}{1-x}\right)^{n-1} \exp\left(\frac{-bx}{1-x}\right)\frac{b}{(1-x)^2}\,dx,$$

since

$$da = \frac{b}{(1-x)^2}\cdot dx.$$

Thus for a given value of b

$$g(x|b)\,dx = \frac{1}{\Gamma(n)}\frac{b^n x^{n-1}}{(1-x)^{n+1}} \exp\left(\frac{-bx}{1-x}\right) dx.$$

Now allow b to vary. The probability that it will lie in the interval db is

$$f(b)\,db = \frac{b^{n-1}e^{-b}}{\Gamma(n)}.$$

Therefore the probability that b will be in the interval db and x in the interval dx simultaneously is

$$\begin{aligned} f(b)\,db\cdot g(x)\,dx &= \frac{1}{[\Gamma(n)]^2}\frac{x^{n-1}}{(1-x)^{n+1}}\left[\int_0^\infty b^{2n-1}\exp\left(\frac{-b}{1-x}\right)db\right]dx \\ &= \frac{1}{[\Gamma(n)]^2}\frac{x^{n-1}}{(1-x)^{n+1}}\frac{\Gamma(2n)}{(1-x)^{-2n}}\,dx. \end{aligned}$$

Then

$$\begin{aligned} g(x) &= \frac{\Gamma(2n)}{[\Gamma(n)]^2}x^{n-1}(1-x)^{n-1}, \quad \text{with} \quad 0 \le x \le 1, \\ &= \frac{x^{n-1}(1-x)^{n-1}}{B(n, n)} \end{aligned}$$

and it is seen that x is a beta variate. Since

$$E(x) = \frac{B(n+1, n)}{B(n, n)} \quad \text{and} \quad E(x^2) = \frac{B(n+2, n)}{B(n, n)},$$

the mean and variance are

$$E(x) = \tfrac{1}{2} \quad \text{and} \quad \operatorname{var}(x) = [4(2n+1)]^{-1}.$$

The distribution tends to normality rapidly with increasing n, and for $n > 50$ we may treat

$$\frac{x - E(x)}{\sqrt{\operatorname{var}(x)}} = 2(x - \tfrac{1}{2})\sqrt{(2n+1)}$$

as a standardized normal variate.

It seems at first sight that this is a test that depends on distance measurements only and that no prior knowledge of the population's density is involved. This is not so, however. In order to choose a random individual from which to measure the distance to its nearest neighbor, the only satisfactory method is to put numbered tags on all the plants in the population and then to consult a random numbers table to decide which of the tagged plants are to be included in the sample. In doing this, we acquire willy-nilly a complete count of the population from which its density automatically follows. There is another method of picking random plants, but it, too, requires that the size of the total population be known. If a sample of size n, say, is wanted from a population of size N, the probability that any given plant in the population will belong to the sample is $p = n/N$. We must then take each population member in turn and decide by some random process whose probability of "success" is p whether that member is to be admitted to the sample. Even if we are willing to guess the magnitude of N intuitively and assign to p a value that will give a sample of approximately the desired size, it is still necessary to subject every population member to a "trial" in order to decide whether it should be included in the sample; as the successive trials are performed, a complete census of the population is automatically obtained.

It must be stressed that it is *not* permissible to take, as a "randomly chosen" member of the population the plant nearest to a random point. This gives a biased sample, a fact to which we shall return.

2. The test proposed by Clark and Evans (1954) requires a knowledge of population density and a sample of n values of r, the distance from a random plant to its nearest neighbor. These distances are not squared.

Let ρ be the number of plants per unit area; that is $\rho = \lambda/\pi$, where λ is the measure of density we have used hitherto, the number of plants per circle of unit radius.

From (10.1) we see that in a randomly dispersed population $E(r) = \frac{1}{2}\sqrt{\rho}$ and from (10.2), that $\text{var}(r) = (4 - \pi)/4\pi\rho$.

Write $\bar{r}$ for the mean of the observed distances. Then, if n, the sample size, is large enough, we may assume that $\bar{r}$ is normally distributed with expectation $\frac{1}{2}\sqrt{\rho}$ and standard error $\sqrt{[(4 - \pi)/4n\pi\rho]}$. This enables us to test for randomness, provided ρ is known exactly. No allowance is made for the sampling variance of an estimated value of ρ.

As an index of nonrandomness, we may use the ratio of the observed to the expected mean distance or

$$R = \frac{\bar{r}}{E(r)} = 2\bar{r}\sqrt{\rho}.$$

Then in a random population $E(R) = 1$; for aggregated populations $R < 1$, since $\bar{r}$ is the mean of *plant*-to-neighbor distances.

3. A third test was described by Pielou (1959) and Mountford (1961). Suppose that we have a sample of n distances measured from random *points* to their nearest plants and that we take the square of the distance, $\omega = r^2$, as variate. Let the population density (in terms of plants per unit area) be ρ. Taking as an index of nonrandomness $\alpha = \bar{\omega}\pi\rho$, we see from (10.4) that $E(\alpha) = 1$ in a population of random pattern. Now suppose that ρ, instead of being exactly known, is estimated. The estimate, denoted by $\hat{\rho}$, is obtained from a sample of m randomly placed quadrats of unit area. Thus $\hat{\rho}$ and $\bar{\omega}$ are both subject to sampling errors. Mountford has shown that

$$E(\alpha) = E(\pi\hat{\rho}\bar{\omega}) = 1$$

and that

$$\operatorname{var}(\alpha) = \frac{1}{n}\left(1 + \frac{n+1}{m\rho}\right).$$

To obtain an estimate of var(α), we must substitute the estimator $\hat{\rho}$ in place of the population value ρ in the last formula.

4. Comparison of the Various Indices of Aggregation

Each of the indices of aggregation we have considered is only a single statistic and can therefore describe only a single aspect of pattern. Each should be thought of as providing only a measure of the extent to which a pattern departs from randomness and nothing more.

A spatial pattern in a continuum obviously has two quite distinct aspects: they may be called *intensity* and *grain*. By the intensity of a pattern we mean the extent to which density varies from place to place. In a pattern of high intensity the differences are pronounced and dense clumps alternate with very sparsely populated zones; when intensity is low, the density contrasts are comparatively slight. The grain of a pattern is independent of its intensity. If the clumps or patches in which the density is relatively high are large in area and widely spaced, we may say that the pattern is coarse-grained. Conversely, if the whole range of different densities is encompassed in a small space, the pattern is fine-grained.

The indices of aggregation calculated from data obtained by sampling with quadrats of one size are all measures of the intensity of a pattern. Lloyd's index of patchiness C and Morisita's index of dispersion I_δ depend on (a) the relative areas occupied by patches of different density (though not on the degree to which these areas are fragmented, which is a question of grain), and (b) on the ratios of the different densities to one another. The V/m ratio,

David and Moore's index of clumping, I, and Lloyd's "mean crowding," $\overset{*}{m}$, depend on (a) and (b) and also on the absolute magnitude of the mean density which does not affect C and I_δ. It must be re-emphasized that all of these indices measure only the intensity, and not the grain, of a pattern. To study grain by means of quadrat sampling it is necessary to use several sizes of quadrats as explained in Chapter 9.

Consider now the indices of aggregation based on distance measurements. Of these, Clark and Evans's index R clearly measures only the intensity of a pattern. Since the distances are measured from plant to plant, most of them will be within-clump distances: the denser the clumps the shorter the measured distances and the smaller the value of R. Thus this index is possibly the best if one particularly wishes to measure pattern intensity. However, it has two defects: it involves very troublesome field work in selecting individuals at random from which to measure distances, and the sampling properties of $\bar{r}$, when the population mean density ρ is estimated from quadrat data, have not yet been worked out.

The two other indices based on distances, Hopkins and Skellam's A and Pielou's and Mountford's α, use point to plant distances. They are therefore influenced by both intensity and grain. This is because some of the random points will fall within high density clumps and the point's nearest neighbor will be a clump member; others will fall where the density of the plants is low, and the nearest one may be in the same sparsely populated patch of ground or in the nearest dense clump.

Separate methods of studying intensity and grain by means of distance measurements have yet to be devised. One possibility, not yet explored, would be to consider not merely the means $\bar{r}$ and $\bar{\omega}$ of the measured distances and their squares but also their variances var(r) and var(ω).

Another possibility has been explored by Morisita (1954) and Thompson (1956). They suggested that more detailed information on a population's pattern could be gained by measuring the distances from random points (or plants) to the nearest, second nearest, third nearest, . . . , neighbors. The method seems to have been little used, probably because the field work is exasperatingly difficult. In a forest, for instance, it is usually quite easy to determine which trees are the nearest and second nearest to a given point; even the third and fourth nearest may be readily recognizable, but the nth neighbor for higher values of n is often difficult to determine. So many trees are approximately equidistant from the point that it becomes almost impossible to rank them or would require such careful measurements that it would not be worth the effort.

It is intuitively apparent, however, that an index of aggregation based on ω_n ($=r_n{}^2$, where r_n is the distance of the nth nearest plant) would decrease with increasing values of n. It would decrease most rapidly in populations with clumps that were small in area, closely spaced, and of few members. In any case, it remains to be discovered what can be learned about a pattern from examination of a curve of index of aggregation versus rank of the neighbor used for its calculation.

The distribution of ω_n when a pattern is random is easily derived. As before, let the mean number of individuals per circle of unit radius be λ. By arguments analogous to those on page 112 we see that the distribution function of ω_n is the probability that a circle of area $\pi\omega_n$ centered on a random point contains at least n individuals; that is

$$F(\omega_n) = 1 - e^{-\lambda\omega_n} - \frac{\lambda\omega_n e^{-\lambda\omega_n}}{1!} - \cdots - \frac{(\lambda\omega_n)^{n-1}e^{-\lambda\omega_n}}{(n-1)!},$$

whence the pdf is

$$f(\omega_n) = F'(\omega_n) = \frac{\lambda^n \omega_n^{n-1} e^{-\lambda\omega_n}}{\Gamma(n)}.$$

The mean* and variance of this distribution are

$$E(\omega_n) = \frac{n}{\lambda} \quad \text{and} \quad \text{var}(\omega_n) = \frac{n^2}{\lambda^2}.$$

This accords with what we should expect intuitively. Comparing this result with (10.4), we see that the expectation of ω_n in a population of density λ is the same as the expectation of ω_1 (the squared distance to the first, or nearest, neighbor) in a population of density λ/n.

5. Sampling Isolated Individuals

It was remarked earlier (page 117) that we cannot obtain a random sample of the individuals in a population by selecting those that are nearest to random points. This procedure gives a biased sample in which relatively isolated individuals are over-represented. To see this, consider a population of just three members, arranged as follows:

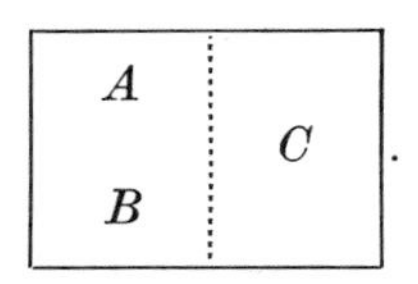

* This is equivalent to asserting that the expected radius^2 of a circle centered at any point and drawn just large enough for the nth nearest individual to the point to lie on the circle's circumference, is n/λ. The expected time to extinction of a population undergoing a pure death process is derived analogously (see page 13). Starting at any instant, if the mean number of deaths is μ per unit of time, the expected time until the ith death occurs is i/μ.

A random point placed in the area has equal probability of falling in the left or right halves, separated by the dashed line. Therefore the isolated individual C is twice as likely as either A or B to be the individual nearest to a random point.

This fact is useful if a sample biased in favor of isolated individuals is what we require. Suppose we want to compare isolated with crowded individuals in order to determine whether they differ in some attribute such as size, age, or susceptibility to disease. Classifying the attribute dichotomously (e.g., big-small, old-young, diseased-healthy) we may compare the proportion of individuals having the attribute in the whole population with the proportion having the attribute in a sample biased in favor of isolated individuals. The biased sample is obtained by taking the nearest individual to each of a number of random points. If the proportions differ significantly, we may conclude that possession of the attribute is related to the degree of isolation of an individual. The method is certainly rather crude, but it obviates the necessity of setting up some arbitrary definition of the "degree of isolation" of an individual and of devising and executing measurements to determine it for each of a large number of individuals (for an example of an application, see Pielou and Foster, 1962).

6. Reciprocity of the Nearest-Neighbor Relation

Whenever a population of individuals is dispersed over a plane, there will be some pairs of individuals such that each is the nearest neighbor of the other; that is, the two individuals are closer to each other than either is to a third individual. For these pairs the nearest-neighbor relationship is reciprocal or reflexive (Clark and Evans, 1955). Given a random pattern, we may calculate the expected number of individuals that are members of reciprocal pairs.

In the diagram that follows, let I_1 and I_2 be two individuals a distance r apart. Then the probability that I_2 is I_1's nearest neighbor is the same as the probability that I_1's nearest neighbor will be at a distance in the range $(r, r + dr)$. This probability is

$$2\pi\rho r e^{-\pi\rho r^2}\, dr = p_1, \text{ say,}$$

where ρ is the density of the population in terms of numbers of individuals per unit area (not per circle of unit radius). From the diagram overleaf we see that the probability that I_2 has I_1 as its nearest neighbor is the probability that the shaded crescent, of area $r^2(\sqrt{3}/2 + \pi/3)$, is empty, and this is

$$p_2 = \exp\left[-\rho r^2\left(\frac{\sqrt{3}}{2} + \frac{\pi}{3}\right)\right].$$

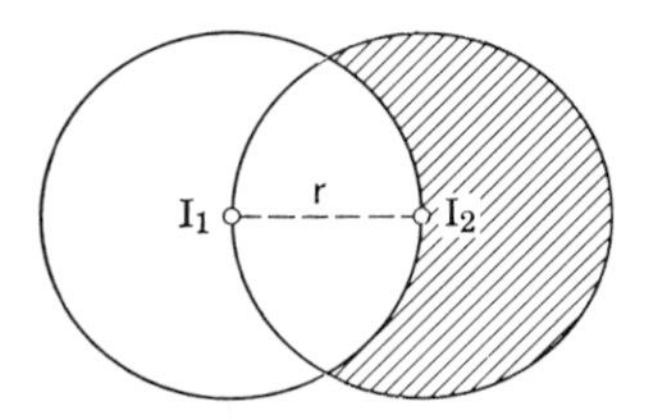

The probability that I_1 will form a reciprocal pair with I_2 (at a distance r apart) is then p_1p_2. Integration over all possible values of r gives the probability P that any individual will be a member of a reciprocal pair.

Thus

$$P = \int_0^\infty 2\pi r\rho \exp\left[-\rho r^2\left(\frac{\sqrt{3}}{2} + \frac{4\pi}{3}\right)\right] dr$$

$$= \frac{6\pi}{3\sqrt{3} + 8\pi} \quad \text{or} \quad 0.6215.$$

The expected proportion of individuals that belong to reciprocal pairs is thus 0.6215 in a population of random pattern. How we might interpret observed departures from expectation is not clear except for the obvious conclusion that an excessively high proportion of reciprocal pairs would support the hypothesis (if it were being tested) that individuals tended to occur as isolated couples. Situations in which such a hypothesis would be worth postulating and in need of testing are hard to think of.

7. The Detection of Regular Spacing

It has so far been assumed that nonrandom patterns always depart from randomness in the direction of aggregation. "Regular" patterns, those in which the individuals are more evenly spaced than they would be in a randomly dispersed population, are easy to visualize but exceedingly rare in nature. This fact is, at first thought, surprising. We might have supposed that in a dense even-aged forest, for example, competition among the trees would cause those that survived beyond the seedling stage to have a regular pattern. Even so, the trees in natural forests are hardly ever found to be regularly arranged. Evidence for regularity has been sought both by quadrat sampling and by using the tests based on distance measurements; the results of these searches have nearly always been negative.

The probable explanation is that, although in very dense parts of a forest short tree-to-neighbor distances may be rare (and distances shorter than a tree's diameter are obviously impossible, since we are assuming that distances are measured from center to center of the individuals), a dearth of short distances may go undetected if the forest as a whole has an aggregated or random pattern. In any comparison of observed and expected values of r or of ω by means of a χ^2-test subdivision of the observed distances into classes is always necessary. If the class containing the smallest variate values is defined to include not only the extremely low values that are impossible or rare but also the moderately low values that are numerous, the rarity of the former will be obscured. Figure 14*a* shows two theoretical probability distributions, A and B, both of which are representable by the same histogram shown in Figure 14*b*. From examination of the histogram we cannot tell whether the distribution of ω is like A or B. Curve B, which is exponential, is what we should expect if the trees' spacing were random and unaffected by competition. Curve A is what would result if competition prevented the occurrence of very short intertree distances. If an empirical distribution, portrayed as a histogram, were like that in Figure 14*b*, we should have no reason to suppose that the parent distribution is like A.

Failure to distinguish between the two kinds of curve will often occur if the data consist of observations on all tree-to-neighbor distances, the long as well as the short. Intertree competition is likely to affect spacing only in localized patches in which the tree density is especially high. To detect it, therefore, it is best to take a sample of short tree-to-neighbor distances only, choosing some arbitrary upper limit for distances to be admitted to the sample. We then have for comparison observed and theoretical distributions both of which are truncated. A method of making the comparison has been described by Pielou (1962*a*).

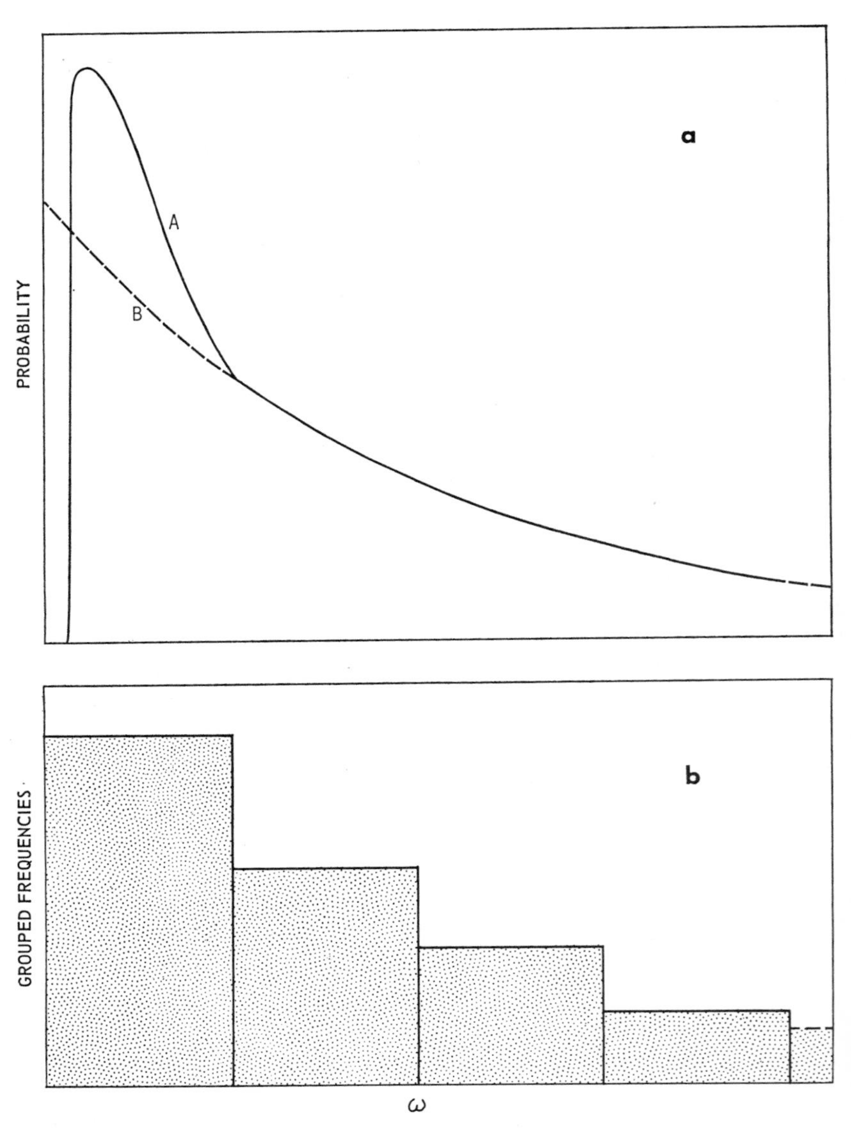

Figure 14*a* Two possible distributions for ω, the square of the distance from a plant to its nearest neighbor. Curve B is exponential. For all but the shortest distances the curves are identical. Figure 14*b*. Histogram corresponding to both curves in Figure 14*a*. The data are coarsely grouped for purposes of illustration.

11

Patterns Resulting from Diffusion

1. Introduction

The discussion of spatial patterns in preceding chapters treated these patterns as static. No consideration has been given to the fact that any pattern must have a history. To reach the locations in which we observe them the organisms of a population must have moved, either actively or passively. The movements of animal populations are of two types: migration and diffusion. Well-known examples of migrations are those of many bird species, caribou herds, breeding seals, locust swarms, butterflies, and the spawning migrations of fishes such as eels and salmon; the list could be greatly extended. Although there is presumably some stochastic element in these mass movements, it seems reasonable to suppose it is overshadowed by the deterministic or "deliberate" element.

In contrast to migration, diffusion* consists in the apparently aimless, undirected movements of animals that seem to be wholly random. For instance, suppose a group of organisms of one species is concentrated at some time $t = 0$ within a small space entirely surrounded by terrain suitable for the species; if we study how the group of organisms spreads out to occupy a larger area, we are studying diffusion. We may consider short-term diffusions, that is, what happens to the original animals within their lifetimes or an even shorter period, or long-term diffusions occupying many generations of the species.

Plant diffusion also invites investigation. Changes in habitat conditions are continually altering the suitability of areas of ground for different species of plants. During the sequence of successional stages at any place the area will become suitable to a succession of different colonizing species that invade

* An alternative word, favored by Odum (1959), for example, is "dispersal." Dispersal is to be contrasted with dispersion which, in ecological contexts, has the same meaning as pattern. Because of the similarity of the words *dispersal* and *dispersion*, it seems preferable to use *diffusion* when movement is meant.

from surrounding areas. More pronounced changes, for example, those wrought by fire, by the drying up of lakes, or by the melting of ice sheets, leave tracts of virgin ground ready to be colonized by plants from neighboring areas.

In short, the locations of animals and plants depend as much on how they got there in the first place as on how they establish themselves and survive once they have arrived. Considering, then, those population shifts that result entirely from random movements, we begin with a discussion of "random walks" and diffusion processes.

2. The Random Walk and Diffusion in One Dimension

Consider a particle that moves in discrete steps, at discrete times, along a line. Suppose at time $t=0$ the particle is at the origin $x=0$. To the left of the origin x takes the values $-1, -2, \ldots,$ and to the right $+1, +2, \ldots$. At each of the times $t=1, 2, \ldots,$ the particle takes a step, to the right with probability p or to the left with probability $q=1-p$.

Let v_{rn} be the probability that after n steps the particle will be at $x=r$. To get there it must have done a total of j steps to the right and $n-j$ steps to the left, with $r=j-(n-j)=2j-n$; then $j=(n+r)/2$ and $n-j=(n-r)/2$. The order of the steps is immaterial. Clearly, $n+r$ must be even, otherwise $v_{rn}=0$.

We see that v_{rn} is a binomial probability; that is

$$v_{rn}=\binom{n}{j}p^j q^{n-j}=\binom{n}{\dfrac{n+r}{2}}p^{(n+r)/2}q^{(n-r)/2}.$$

Here $\binom{n}{j}$ is the number of ways in which the particle can reach $x=r$ in n steps.

Now let us pass to the limit, allowing both the step lengths and the time intervals to become infinitesimally small. First, let each step be of length Δx. We can now find the mean and variance of the displacement *per step*. The mean displacement will be positive if the net displacement is to the right of the origin and negative if to the left. We write the two possible events and their probabilities in the form of Table 7 and determine their mean and variance in the usual way.

TABLE 7

DISPLACEMENT i	PROBABILITY π	πi	πi^2
$+\Delta x$	p	$p\,\Delta x$	$p(\Delta x)^2$
$-\Delta x$	q	$-q\,\Delta x$	$q(\Delta x)^2$

Then the mean is $m = (p - q)\,\Delta x$ and the variance is

$$\begin{aligned}\sigma^2 &= (p+q)(\Delta x)^2 - (p-q)^2(\Delta x)^2 \\ &= 4pq(\Delta x)^2.\end{aligned}$$

Suppose now that the time interval between steps is of length Δt. In a period of length t approximately $t/\Delta t$ steps will be taken (or exactly $t/\Delta t$ if t is a multiple of Δt). Then the total displacement in time t has mean and variance given to a very close approximation by

$$m_t = \frac{t}{\Delta t}(p-q)\,\Delta x \quad \text{and} \quad \sigma_t{}^2 = \frac{t}{\Delta t}\cdot 4pq(\Delta x)^2.$$

Now let both $\Delta x \to 0$ and $\Delta t \to 0$. At the same time $\Delta x/\Delta t$ and also $(\Delta x)^2/\Delta t$ must be allowed to take suitably chosen values; otherwise nonsensical results will be obtained. Note that since p and q, the probabilities of steps to right and left, do not change with time and the steps occur at equal intervals the expected displacement must be proportional to the time elapsed. So we may put $m_t = 2ct$ where $2c$ is a constant of proportionality. Likewise, since the steps are independent, $\sigma_t{}^2$, the variance of the displacement at time t, is the sum of the variances pertaining to each step. Thus $\sigma_t{}^2$ is also proportional to t and we may write $\sigma_t{}^2 = 2Dt$ where $2D$ is another constant of proportionality. This is equivalent to putting

$$2c = (p-q)\frac{\Delta x}{\Delta t} \quad \text{and} \quad 2D = \frac{4pq(\Delta x)^2}{\Delta t}.$$

Thus we must allow Δx and Δt to tend to zero in such a way that c and D, hence the mean and variance of the displacement, remain finite. This is equivalent to requiring that the x- and t-scales be in appropriate ratio to each other. For the mean to remain finite it is also necessary that $p - q$ be small, of the same order of magnitude as Δx; that is, we must have $p - q = O(\Delta x)$. Then

$$4pq = 1 - (p-q)^2 = 1 - [O(\Delta x)]^2 \to 1 \quad \text{as} \quad \Delta x \to 0,$$

hence

$$\frac{(\Delta x)^2}{\Delta t} = 2D.$$

Also $p - q = c\Delta x/D$ and $p + q = 1$ and we have

$$p = \frac{1}{2} + \frac{c}{2D}\Delta x \quad \text{and} \quad q = \frac{1}{2} - \frac{c}{2D}\Delta x.$$

The two constants introduced, namely c and D, are known as the coefficients of *drift* and *diffusion*, respectively.

It will be recalled that v_{rn}, the probability that a particle will be at position r at time n, is a binomial probability. As the steps become short and numerous, this probability may therefore be approximated by the normal probability function with mean $2ct$ and variance $2Dt$; that is,

$$\phi(x, t)\, dx = \frac{1}{\sqrt{4\pi Dt}} \exp\left[-\frac{1}{4Dt}(x-2ct)^2\right] dx.$$

is the probability that at time t, the particle's position will be in the interval $(x - \frac{1}{2}\, dx,\ x + \frac{1}{2}\, dx)$. If there is no drift, that is, if $c = 0$,

$$\phi(x, t) = \frac{1}{\sqrt{4\pi Dt}} \exp\left(\frac{-x^2}{4Dt}\right). \tag{11.1}$$

Starting from the same premises, we shall now derive the partial differential equation that describes diffusion with drift (the Fokker-Planck equation of physics), and we shall also show that (11.1) is a solution of it when the drift is zero.

Write $\phi(x, t)$ for the probability that at time t the particle will be at position x. We can immediately write down the difference equation

$$\phi(x, t + \Delta t) = p\, \phi(x - \Delta x, t) + q\, \phi(x + \Delta x, t),$$

since for the particle to be at x at time $t + \Delta t$ it must have been at either $x - \Delta x$ or $x + \Delta x$ at time t.

Expanding both sides by Taylor's theorem and writing ϕ for $\phi(x, t)$ gives

$$\phi + \frac{\partial \phi}{\partial t} \cdot \Delta t + \frac{1}{2!}\frac{\partial^2 \phi}{\partial t^2}(\Delta t)^2 + \cdots = \phi + (q - p)\frac{\partial \phi}{\partial x} \cdot \Delta x + \frac{1}{2!}\frac{\partial^2 \phi}{\partial x^2}(\Delta x)^2 + \cdots$$

or, on dividing through by Δt,

$$\frac{\partial \phi}{\partial t} + \frac{1}{2!}\frac{\partial^2 \phi}{\partial t^2} \cdot \Delta t + \cdots = (q - p)\frac{\partial \phi}{\partial x} \cdot \frac{\Delta x}{\Delta t} + \frac{1}{2!}\frac{\partial^2 \phi}{\partial x^2}\frac{(\Delta x)^2}{\Delta t} + \cdots.$$

As before let Δx and $\Delta t \to 0$ in such a way that $(p - q)\, \Delta x/\Delta t \to 2c$ and $(\Delta x)^2/\Delta t \to 2D$, whereas Δt and its powers (on the left-hand side) and $(\Delta x)^3/\Delta t$ and higher terms (on the right-hand side) tend to zero. Then

$$\frac{\partial \phi}{\partial t} = -2c\frac{\partial \phi}{\partial x} + D\frac{\partial^2 \phi}{\partial x^2}.$$

This is the Fokker-Planck equation for one-dimensional diffusion with drift.

If there is no drift so that $c = 0$,

$$\frac{\partial \phi}{\partial t} = D\frac{\partial^2 \phi}{\partial x^2}. \tag{11.2}$$

We now show that the normal probability function with mean zero, as given in (11.1), is a solution of (11.2). From (11.1)

$$\frac{\partial \phi}{\partial t} = \frac{e^{-x^2/4Dt}}{4\sqrt{\pi D t^3}}\left(\frac{x^2}{2Dt} - 1\right)$$

and

$$\frac{\partial^2 \phi}{\partial x^2} = \frac{e^{-x^2/4Dt}}{4\sqrt{\pi D^3 t^3}}\left(\frac{x^2}{2Dt} - 1\right),$$

so that $\partial\phi/\partial t = D(\partial^2\phi/\partial x^2)$ as required.

So far we have considered only diffusion in one dimension; in the case of diffusion in the plane the diffusion equation is

$$\frac{\partial \phi}{\partial t} = D\,\nabla^2\phi,$$

where

$$\nabla^2 = \frac{\partial^2}{\partial x^2} + \frac{\partial^2}{\partial y^2}.$$

The required solution is the joint normal distribution

$$\phi(x, y, t) = \frac{1}{4\pi D t}\exp\left[\frac{-(x^2+y^2)}{4Dt}\right], \tag{11.3}$$

with $\text{var}(x) = \text{var}(y) = 2Dt$ as before.

3. Alternative Derivation of the Two-Dimensional Diffusion Equation

In ecological contexts we are far more often concerned with two dimensions than with one. It is therefore worthwhile to derive the two-dimensional diffusion equation *de novo* in the way shown by Skellam (1951), for instance.

Imagine that a particle that can move in any direction in the plane is displaced through a distance ε at times $t, t + \Delta t, t + 2\,\Delta t, \ldots$. Then at any moment $t + \Delta t$ it must lie somewhere on a circle of radius ε centered on the position that it occupied at time t. Therefore $\phi(x, y, t + \Delta t)$, the probability density at time $t + \Delta t$ at the point (x, y) is the mean of $\phi(\xi, \eta, t)$ over all the points (ξ, η) on a circle of radius ε with center (x, y). Thus

$$\phi(x, y, t + \Delta t) = \frac{1}{2\pi}\int_0^{2\pi} \phi(\xi, \eta, t)\, d\theta.$$

Putting $\xi = x + \varepsilon\cos\theta$ and $\eta = y + \varepsilon\sin\theta$, it is seen that

$$\phi(x, y, t + \Delta t) = \frac{1}{2\pi}\int_0^{2\pi} \phi(x + \varepsilon\cos\theta, y + \varepsilon\sin\theta, t)\, d\theta. \tag{11.4}$$

Expanding the left-hand side in the form of a Taylor's series and writing ϕ for $\phi(x, y, t)$ gives

$$\phi(x, y, t + \Delta t) = \phi + \frac{\partial \phi}{\partial t} \cdot \Delta t + \frac{1}{2!} \frac{\partial^2 \phi}{\partial t^2} (\Delta t)^2 + \cdots.$$

Similarly, expanding the right-hand side gives

$$\frac{1}{2\pi} \int_0^{2\pi} \Bigg[\phi + \varepsilon\left(\cos\theta \cdot \frac{\partial \phi}{\partial x} + \sin\theta \cdot \frac{\partial \phi}{\partial y}\right)$$

$$+ \frac{\varepsilon^2}{2!}\left(\cos^2\theta \frac{\partial^2 \phi}{\partial x^2} + 2\cos\theta\sin\theta \frac{\partial^2 \phi}{\partial x\, \partial y} + \sin^2\theta \frac{\partial^2 \phi}{\partial y^2}\right) + \cdots\Bigg] d\theta,$$

but, since

$$\int_0^{2\pi} \cos\theta\, d\theta = \int_0^{2\pi} \sin\theta\, d\theta = \int_0^{2\pi} \cos\theta \sin\theta\, d\theta = 0$$

and $\displaystyle\int_0^{2\pi} \cos^2\theta\, d\theta = \int_0^{2\pi} \sin^2\theta\, d\theta = \pi,$

this reduces to

$$\phi + \frac{\varepsilon^2}{2!} \cdot \frac{1}{2\pi}\left(\pi \frac{\partial^2 \phi}{\partial x^2} + \pi \frac{\partial^2 \phi}{\partial y^2}\right).$$

Equation 11.4 may now be written as

$$\frac{\partial \phi}{\partial t} + \frac{1}{2!} \frac{\partial^2 \phi}{\partial t^2} \Delta t + \cdots = \frac{\varepsilon^2}{4\,\Delta t}\left(\frac{\partial^2 \phi}{\partial x^2} + \frac{\partial^2 \phi}{\partial y^2}\right) + \text{negligible terms in } \varepsilon^3/\Delta t$$

Therefore

$$\frac{\partial \phi}{\partial t} = D\,\nabla^2 \phi, \tag{11.5}$$

where $D = \varepsilon^2/4\Delta t$.

Here ε is the distance in the *plane* covered by an infinitesimal step of the particle, so that $\varepsilon^2 = (\Delta x)^2 + (\Delta y)^2$. Thus, if, as before, we put $(\Delta x)^2/\Delta t = 2D$, we have, analogously $\varepsilon^2/\Delta t = 4D$. Considerations of symmetry show that $\text{var}(x) = \text{var}(y) = 2Dt$. Thus the joint distribution of x and y, the coordinates of the particle at time t, has pdf

$$\phi(x, y, t) = \frac{1}{4\pi Dt} \exp\left[\frac{-(x^2 + y^2)}{4Dt}\right]$$

and this will be seen to be a solution of (11.5).

We now consider three applications of diffusion theory to ecological problems. In all three examples it is assumed that diffusion occurs without drift.

4. The Rate of Spread of a Population Over a Plane

We have hitherto considered the probability, given by $\phi(x, y, t)\,dx\,dy$, that a particular particle starting at the origin at time $t=0$ will, at a later time t, be found in the element of area having coordinates in the ranges $x \pm \frac{1}{2}\,dx$ and $y \pm \frac{1}{2}\,dy$. If, instead of dealing with a single particle, we suppose that at $t=0$ a whole population of particles had been concentrated at the origin, then $\phi(x, y, t)\,dx\,dy$ denotes the proportion of the population to be expected in this element of area at time t. The distribution is that of shots around a bulls-eye.

Since what interests us is the way in which density falls off with distance from the center of diffusion, it is convenient to convert to polar coordinates. Then

$$x = r\cos\theta; \qquad y = r\sin\theta; \qquad dx\,dy = r\,dr\,d\theta,$$

and

$$\phi(r, \theta, t)\,dr\,d\theta = \frac{r}{4\pi Dt}\exp\left(\frac{-r^2}{4Dt}\right)dr\,d\theta.$$

Following Skellam (1951), we now write $4D = a^2$. Thus a^2 is the mean-square displacement in a unit of time. Then

$$\phi(r, \theta, t)\,dr\,d\theta = \frac{r}{\pi a^2 t}\exp\left(\frac{-r^2}{a^2 t}\right)dr\,d\theta.$$

Integrating over θ gives $\phi(r, t)\,dr$, the expected proportion of the population of particles whose distance from the origin lies in the range $r \pm \frac{1}{2}\,dr$. Thus

$$\phi(r, t) = \int_0^{2\pi} \frac{r}{\pi a^2 t}\exp\left(\frac{-r^2}{a^2 t}\right)d\theta = \frac{2r}{a^2 t}\exp\left(\frac{-r^2}{a^2 t}\right). \tag{11.6}$$

We now wish to determine the rate at which a diffusing population spreads. Clearly, if there is no drift, the contours of equal density will move out in ever-expanding circles like ripples on a pond. To find the expected size of the boundary circle, which contains the whole population at time t, we evaluate p_t, the proportion of the population that, at time t, is expected to be farther than a distance R_t from the center of diffusion. Obviously,

$$p_t = \int_{R_t}^{\infty} \frac{2r}{a^2 t}\exp\left(\frac{-r^2}{a^2 t}\right)dr = \exp\left(\frac{-R_t^2}{a^2 t}\right).$$

Let the number of particles in the population be N and put $p_t = 1/N$. This amounts to choosing a value of p_t, hence of R_t, such that only one member of the population is expected to be farther than R_t from the origin or that all but one member of the whole population is contained in a circle of radius R_t. Then

$$\frac{1}{N} = \exp\left(\frac{-R_t^2}{a^2 t}\right)$$

or

$$R_t^2 = a^2 t \ln N.$$

The area of the circle containing the expanding population is thus proportional to the time elapsed since diffusion started. N is assumed to remain constant; therefore the density of the population must perforce get progressively less as time passes.

In the foregoing it was assumed that the diffusing organisms did not reproduce or die so that N did not vary with time. We next explore what happens if the particles are breeding organisms and observation is continued over many generations. Suppose the population, initially consisting of N_0 individuals, is increasing exponentially as the result of a simple birth and death process. Then at time t the size of the population is $N_t = N_0 e^{ct}$, where c is the intrinsic rate of natural increase (see Chapter 1).

Therefore $\ln N_t = ct +$ constant, and for sufficiently small N_0 we may take $\ln N_t = ct$. The radius at time t of the circle containing the population, which is expanding both numerically and spatially, is then given by

$$R_t^2 = a^2 t \ln N_t = a^2 t^2 c.$$

We see that in this case the radius (rather than the area) of the circle containing the population is proportional to t.

Skellam (1951) examined the rate of spread of muskrats which were introduced into Central Europe in 1905. On five occasions in the succeeding 23 years the area occupied by the increasing population was mapped, thus enabling contours to be drawn to show the population's extent at five different times. As we should expect, the contours are not circular, but there is no evidence of drift, and it seems reasonable to treat the area within each contour as an estimate of πR_t^2. If we assume, further, that the muskrat population was increasing in numbers exponentially, as could well happen with an immigrant population spreading into areas not occupied by potential competitors, we should expect the square root of the area within each contour to be linearly related to time. Skellam found that this relationship did, in fact, exist; in a plot of $\sqrt{\text{area}}$ versus time the five points lie very close to a straight line passing through the origin.

Skellam also considered the rate of northward spread of oak trees in Great Britain after the melting of the last Pleistocene ice sheet. Assuming, very roughly, that the oaks advanced about 600 miles in 20,000 years, he relates this inferred rate of advance with (a) the presumed rate of numerical increase of the population of oaks and (b) the distance to which the acorns from a parent oak are disseminated. He concluded that (unless some oaks survived at sheltered unglaciated spots within the ice field) the speed of reinvasion of the oaks could have resulted only from acorns being transported by animals; the advance was far too rapid for one to assume that acorns fell to the ground close to their parent trees and germinated where they landed.

This last result is an excellent demonstration of the usefulness of mathematical models in ecology. Even though one would rarely expect such highly abstract arguments as those leading to the diffusion equations to be directly applicable to the complicated behavior of living organisms, they do permit one to reach convincing conclusions about the rates of spread that are possible in different circumstances. For many of the plant and animal species that invade hitherto unoccupied areas it would be interesting to know whether their spread has been purely passive or was accelerated by some extrinsic agency.

5. A Spatial Pattern Resulting from a Diffusion Process

Suppose an insect lays a compact cluster of eggs and that after hatching the larvae diffuse outward. Then, as already given in (11.6), the expected proportion of them in the range $r \pm \frac{1}{2}\,dr$ at time t is

$$\phi(r, t)\,dr = \frac{2r}{a^2 t} \exp\left(\frac{-r^2}{a^2 t}\right) dr.$$

Next assume that after moving (i.e., performing a random walk in the plane) for a time a larva stops and remains at the spot it has reached. For each larva the probability that it will stop in any short time interval Δt is $\lambda\,\Delta t$, with λ a constant. Thus the travel times for the larvae are independent random variates with pdf $\pi(t) = \lambda e^{-\lambda t}$. (This is the pdf of the intervals between independent events occurring at a mean rate of λ per unit of time; the derivation is given on page 25.)

Then the distribution of the distances of the larvae from the center of diffusion after all movement has ceased is given by

$$\begin{aligned} g(r)\,dr &= \left[\int_0^\infty \phi(r, t)\pi(t)\,dt\right] dr \\ &= \left[\int_0^\infty \frac{2r}{a^2 t}\lambda \exp\left(-\lambda t - \frac{r^2}{a^2 t}\right) dt\right] dr. \end{aligned}$$

This integral may be simplified. Put $\rho = 2r\sqrt{\lambda}/a$ so that $d\rho = 2\sqrt{\lambda}dr/a$. Then, since

$$g(r)\,dr = \left[\frac{1}{2}\int_0^\infty \frac{2r\sqrt{\lambda}}{a}\exp\left(-\lambda t - \frac{r^2}{a^2 t}\right)\frac{dt}{t}\right]\frac{2\sqrt{\lambda}}{a}\,dr,$$

we have

$$h(\rho)\,d\rho = \rho\left[\frac{1}{2}\int_0^\infty \exp\left(-\lambda t - \frac{\rho^2}{4\lambda t}\right)\frac{dt}{t}\right]d\rho,$$

where $h(\rho)$ is the pdf of ρ. Next put $\lambda t = \tau$. Then $\lambda\,dt = d\tau$ and $dt/t = d\tau/\tau$. Thus

$$h(\rho)\,d\rho = \rho\left[\frac{1}{2}\int_0^\infty \exp\left(-\tau - \frac{\rho^2}{4\tau}\right)\frac{d\tau}{\tau}\right]d\rho.$$

The integral in brackets is tabulated. It is $K_0(\rho)$, a modified Bessel function of the second kind, of which tables will be found in, for instance, Watson (1944). Therefore

$$h(\rho) = \rho\,K_0(\rho) = \frac{2r\sqrt{\lambda}}{a}\,K_0\left(\frac{2r\sqrt{\lambda}}{a}\right);$$

that is, the expected number of larvae in the annulus between circles of radius ρ and $\rho + d\rho$ is proportional to $\rho\,K_0(\rho)\,d\rho$. The area of this annulus is $2\pi\rho\,d\rho$; hence the expected number of larvae per unit area at distance ρ from the center of diffusion is proportional to $K_0(\rho)$. The pattern of the larvae is radially symmetrical, and the way in which density falls off with distance is shown in Figure 15. This is simply the curve of $K_0(\rho)$ versus ρ; [e.g., see Wylie (1951)]. Different curves of this form differ among themselves only in the relative scales of ordinate and abscissa or, equivalently, in the scale factor $2\sqrt{\lambda}/a$. Therefore the shape of the distribution depends only on the ratio $2\sqrt{\lambda}/a$ and not on the magnitudes of the separate components $\sqrt{\lambda}$ and a; λ is the reciprocal of the expected travel time of a larva and a^2 is the mean-square displacement in unit time, given that no stopping occurs (see page 131). Thus we see that it would be possible for two species conforming to this model to have identical patterns if one of them diffused slowly and had a long mean travel time, whereas the other diffused rapidly and had a short mean travel time.

Broadbent and Kendall (1953) consider the applicability of this distribution to the pattern of the larvae of helminths parasitic on sheep and rabbits. The eggs are present in numbers in the excreta of the hosts and, after hatching, diffuse outward through the grass of a pasture. After an interval t [presumed

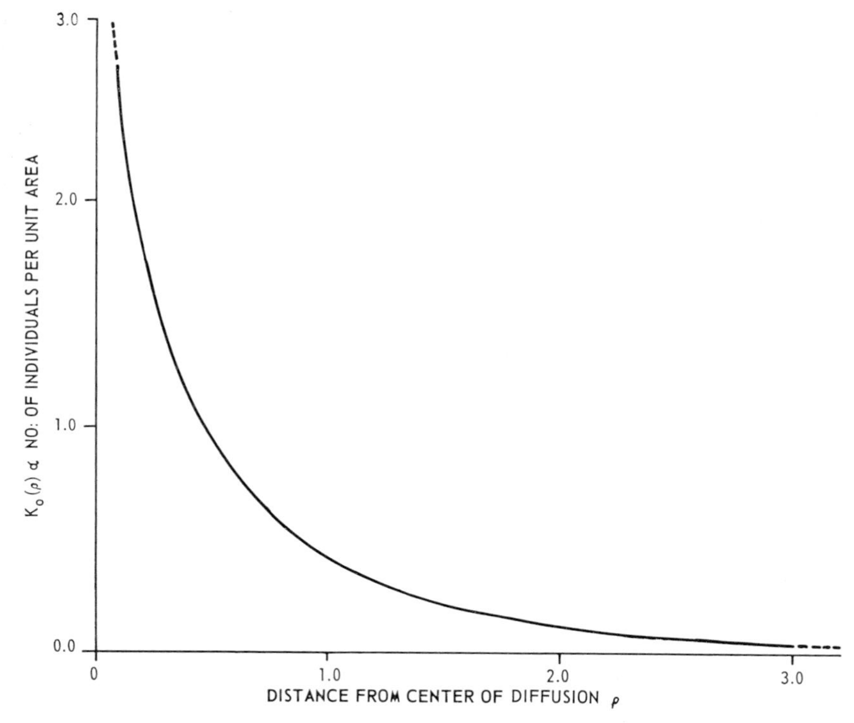

Figure 15. The relationship between density and distance from point of diffusion when organisms diffuse outward from a center and stop at random times.

to have distribution $\pi(t)$] each larva climbs a grass blade and stays there until ingested by a grazing animal, when the cycle starts again.

Williams (1961) discusses the appropriateness of the Bessel function distribution for describing the pattern of codling moth larvae in an apple orchard, where a large number of adult moths had been released at a central point. He gives a method of estimating the distribution's single parameter, the scale factor $2\sqrt{\lambda/a}$.

6. The Probability of Reaching a Specified Destination and the Time Taken to Reach It

In this section we shall consider, for simplicity, only one-dimensional diffusion. Examples of ecological contexts in which the one-dimensional theory might apply are the movements of shore-dwelling animals or of aquatic animals in narrow watercourses (presumed to be stagnant so that the possibility of drift may be ignored).

Suppose a group of organisms is diffusing from a starting point at a distance x from a destination which we shall call "home." If they reach it, they stop. Two of the questions that may be asked (and, provided our assumptions hold, answered) are (a) what proportion of the starting population will have reached home by time t_0, assuming diffusion to have started at $t = 0$? And (b) what is the average speed of return of those that do reach home?

Consider first the random-walk model (see page 126) in which a particle is assumed to take discrete steps of constant length. For simplicity we also assume that there is no drift or, in other words, that steps to right and left are equiprobable. We now ask: what is the probability that the particle will reach "home," at a distance r steps to the right of the starting point, on the nth step *for the first time*? Let this probability be u_{rn}. It will now be proved that

$$u_{rn} = \frac{r}{n} \binom{n}{\frac{n+r}{2}} \frac{1}{2^n}, \tag{11.7}$$

with $u_{rn} = 0$ when $n + r$ is odd.

Figure 16 shows the set-up diagramatically. The starting point is at the origin A, with coordinates (0, 0). At every step the particle moves one unit

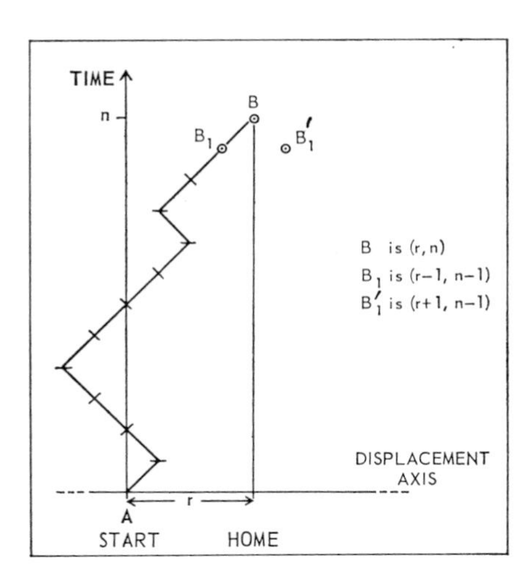

Figure 16. To illustrate the derivation of u_{rn}, the probability that a particle starting at 0 will reach "home" at a distance r for the first time on the nth step (see text).

upward, along the time axis, and one unit to right or left, toward or away from home. Any sequence of steps taken by the particle may therefore be plotted as a path on the graph like the one shown. If the particle reaches home on the nth step, the path must pass through B which has coordinates (r, n). For it to arrive there on the nth step *for the first time* the whole path must lie to the left of (must not touch or cross) the vertical line through B. Calling such a path an "admissible path," it follows that

$$u_{rn} = \frac{\text{number of admissible paths}}{\text{total number of possible paths with } n \text{ steps}}.$$

Now, since at every step the particle has two possibilities (it may go right or left) the total number of possible paths is 2^n. It remains to determine the number of admissible paths, say N^*_{AB}. This may be done as follows:

Note first that all admissible paths must go through B_1 with coordinates $(r-1, n-1)$.

Put $N(AB_1)$ = total number of paths from A to B_1.

$N_L(AB_1)$ = number of paths from A to B_1 that are wholly to the left of the vertical through B. (*Note.* This is B, not B_1.) These paths are "admissible."

$N_R(AB_1)$ = number of paths from A to B_1 that touch or cross the vertical through B. These paths are "inadmissible."

Then

$$N(AB_1) = N_L(AB_1) + N_R(AB_1).$$

and also

$$N^*_{AB} = N_L(AB_1).$$

Next, consider the point B_1' at $(r+1, n-1)$, which is the reflection of B_1 in the vertical through B. Contemplation of the figure will show that

$$N_R(AB_1) = N(AB_1'),$$

that is, the number of *in*admissible paths from A to B_1 is equal to the total number of paths by any route whatever from A to B_1'. This follows, since inadmissible paths to B_1 must reach the vertical through B at some point (r, m), say, with $m \leq n-2$. Thereafter, from symmetry, the number of paths to B_1 is the same as the number of paths to B_1'. Therefore

$$\begin{aligned} N^*_{AB} &= N_L(AB_1) = N(AB_1) - N_R(AB_1) \\ &= N(AB_1) - N(AB_1'). \end{aligned}$$

Recalling that the total number of paths from the origin to the point (r, n) is $\binom{n}{(n+r)/2}$ and consequently that

$$N(AB_1) = \binom{n-1}{\frac{n+r-2}{2}} \quad \text{and} \quad N(AB_1') = \binom{n-1}{\frac{n+r}{2}},$$

it finally follows that

$$N^*_{AB} = \binom{n-1}{\frac{n+r}{2}-1} - \binom{n-1}{\frac{n+r}{2}} = \frac{r}{n}\binom{n}{\frac{n+r}{2}},$$

whence

$$u_{rn} = \frac{N^*_{AB}}{2^n} = \frac{r}{n}\binom{n}{\frac{n+r}{2}}\frac{1}{2^n}$$

as in (11.7). Here

$$\binom{n}{\frac{n+r}{2}} = 0 \quad \text{if} \quad n+r \text{ is odd.}$$

An entirely different proof will be found in Bailey (1964). It is the solution of the famous Gambler's Ruin problem of probability theory: if a gambler with capital of $\$r$, playing against an infinitely rich adversary, bets \$1 on the outcome of each of a series of tosses with a fair coin, u_{rn} is the probability that he will be ruined at the nth toss.

We now pass to the limit. Denote the step lengths by Δx and the time intervals by Δt and allow $\Delta x, \Delta t \to 0$. Recalling that when $p = q = \frac{1}{2}$ the limiting form of

$$v_{rn} = \binom{n}{\frac{n+r}{2}}\frac{1}{2^n}$$

is

$$\phi(x, t) = \frac{1}{2\sqrt{\pi D t}}\, e^{-x^2/4Dt} \qquad \text{[see (11.1)]},$$

we see by analogy that the limiting form of $u_{rn} = (r/n)v_{rn}$ is

$$\psi(x, t) = \frac{x}{t}\,\frac{1}{2\sqrt{\pi D t}}\, e^{-x^2/4Dt}.$$

[For a rigorous proof of this, see Feller (1968).] That is, the probability that the particle will, for the first time, reach "home" at a distance x from its starting point in the interval $t \pm \frac{1}{2}\, dt$ is $\psi(x, t)\, dt$.

We return now to the ecological questions posed at the beginning of this section. From the above arguments we may conclude that if a group of organisms is released at time $t = 0$ at a distance x from home and then diffuses in one dimension the expected proportion of the group that will reach home on or before time t_0 is

$$Q(t_0, x) = \int_0^{t_0} \psi(x, t)\, dt.$$

The average speed of those that do reach home can also be derived. Thus an organism that travels a distance x in time t has traveled at an average speed of x/t. Therefore the mean speed of all the organisms that reach home (i.e., excluding those that have not arrived by time t_0) is

$$\bar{s}(t_0, x) = \frac{\int_0^{t_0} (x/t)\psi(x, t)\, dt}{Q(t_0, x)}.$$

These results are due to Wilkinson (1952) who applied them to experimental studies on the homing ability of sea birds. The birds were taken along a coast to a distance x from home and then released to find their way back. Observational results included records of (a) the percentage of birds that was successful in returning home before a given time had elapsed and (b) the average speed of the birds that did reach home within this time. Wilkinson found that the observations accorded well with what we should expect on the hypothesis that the homing movements of the birds amounted to random diffusion in one dimension (i.e., along the coast). We could infer (though Wilkinson does not) that the homing birds had no navigational ability and that their movements were wholly random. What Wilkinson does conclude is this: that *if* the birds have navigational ability more detailed observations than those at his disposal will be needed to reveal it.

12

The Patterns of Ecological Maps: Two-Phase Mosaics

1. Introduction

In this chapter we return to a consideration of ecological patterns from the static point of view and discuss the third of the three types described on p. 81. These "case 3" patterns are exhibited by vegetatively reproducing plants, which commonly occur as extensive clumps of shoots. Regardless of whether the individual shoots of a clump are densely packed or well separated, it is natural to treat the clumps rather than the shoots as the entities whose pattern is to be studied. A clump need not be a distinct, separate unit, nor need it be an "individual" in the genetic sense.

When the pattern of a clumped species can be mapped—and this can be done only if the clump boundaries are recognizable—the result is a two-phase mosaic with a patch-phase (where the plant occurs) and a gap-phase (where it is absent). The patches are of indefinite size and shape and lack definable centers. They do not necessarily constitute "islands" in the gap-phase; in some parts of an area the patch-phase may form a continuum and the gap-phase occur as lacunae. Both maps in Figure 12 show two-phase mosaics; in them the patch-phase is shown as occupied by separate but fairly closely spaced shoots. Many species of plants, for instance those that spread by long stolons or have rhizomes from which the above-ground shoots grow at wide intervals, do not present such easily distinguishable patch- and gap-phases. However, we defer a discussion of such plants to page 153 and confine our attention for the moment to species whose patterns can be easily mapped as mosaics. In this chapter only two-phase mosaics, obtained by mapping a single plant species, are discussed. Mosaics with more than two phases are considered in Chapter 16. It will be noticed that other kinds of mosaics, besides those showing the patches of the different species in tracts of vegetation, are often studied by ecologists. Examples are maps or aerial photos showing land versus water or forest versus grassland; also soil, physiographic, and geological maps. Thus the properties of mosaic maps are of interest in a wide range of contexts and merit detailed consideration.

It is clear that two-phase mosaic maps are entirely different from the "dot maps" that can be used satisfactorily to portray the patterns of species occurring as distinct, widely scattered individual plants, and it is worth inquiring what properties such a mosaic must have to be regarded as "random". There is no unique answer, as there is in the case of a dot map. A dot map is random if and only if the locations of the dots, which are treated as dimensionless, form a realization of a Poisson point process in the plane; that is, if every point in an area is as likely as every other to be the location of a dot and the dots are independent of one another. The number of dots in a small sample area or quadrat is then a Poisson variate. For two-phase mosaics, on the other hand, no such unambiguous definition of randomness is possible. There are various ways in which a random mosaic may be constructed, and we cannot define a mosaic as random without also specifying exactly what sort of randomness is meant. We now discuss two kinds of random mosaics which, for brevity, are here named L-mosaics and S-mosaics.

2. The Random-Lines Mosaic, or L-Mosaic

Suppose that in the map of a mosaic, the patch-phase is to be colored black and the gap-phase left white. One way of constructing a random mosaic is to draw a set of "random lines" that subdivide an area into a network of convex polygons or "cells." Each cell is then independently assigned its color with fixed probabilities, say b for a black cell and w for a white cell with $b + w = 1$. When contiguous cells receive the same color, they form a many-celled patch, if black, or a many-celled gap, if white. In the arguments that follow it is important to keep clear the distinction between a cell and a patch or gap. The cells, which are always convex, are the small areas formed when the random lines are drawn across the area and are the units of which the patches and gaps are composed. A patch consists of any number of contiguous cells that chance to be colored black; likewise, any number of contiguous white cells constitutes a gap.

A method of drawing random lines is as follows: suppose the area in which the lines are to be drawn is circumscribed by a circle of radius r. Take the center of this circle as the pole of a polar coordinate frame and draw an initial line through it. Now, from a random-numbers table take pairs of random polar coordinates, (p, θ), say, with p in $(0, r)$ and θ in $[0, 2\pi)$. For each such coordinate pair, for example (p_1, θ_1), a line may be drawn which passes through the point with coordinates (p_1, θ_1) and is perpendicular to the line joining (p_1, θ_1) and the pole of the coordinates; this is a random line.

Most of these random lines will lie across the map area, though a few may fall entirely outside it. Those that cross the area will subdivide it into a network of convex polygonal cells. These are the cells that are to be colored black or white. Figure 17 shows a mosaic prepared in this way. It is convenient to call it a random-lines mosaic or an L-mosaic for short.

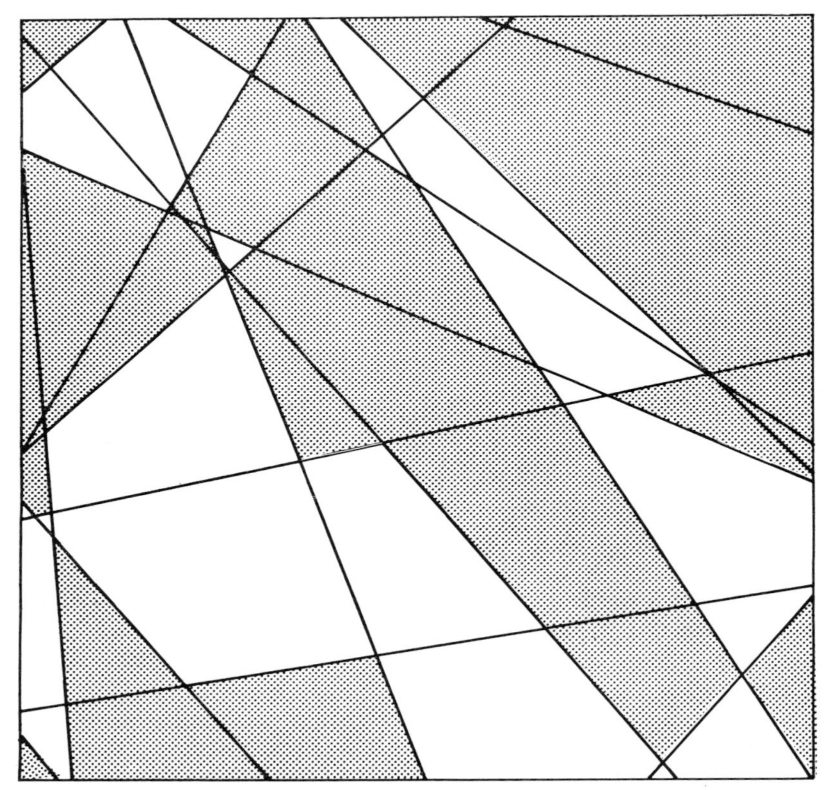

Figure 17. A random-lines mosaic or "L-mosaic."

Now suppose that such a mosaic is sampled at equidistant points along a line transect and that the color at each point (B or W) is recorded. The observations will be a sequence such as BBB WW BB WWW B WWW BB ..., for example. We shall now show that for a mosaic constructed in the manner just described, that is, an L-mosaic, this sequence constitutes a realization of a simple two-state Markov chain. To do this we must show that the probability that the ith point will be a B, say, depends only on the color at the $(i-1)$th point and is independent of the colors at earlier points, the

$(i-2)$th, $(i-3)$th, and so on. This must hold for all i. A rigorous proof has been given by Switzer (1965) and here we are content with a heuristic approach.

Let $p(B_i, B_{i+1})$, for example, be the probability that both the ith point and the $(i+1)$th point on the transect will be B's. Denote by $\pi_{i,i+1}$ the probability that these two adjacent transect points will lie in the same *cell* of the random-lines network. Then

$$p(B_i, B_{i+1}) = \pi_{i,i+1}b + (1 - \pi_{i,i+1})b^2; \tag{12.1}$$

that is, $p(B_i, B_{i+1})$ is the sum of two terms. The first, $\pi_{i,i+1}b$, is the probability that both points will be in the same cell and that the cell is black. The second term, $(1-\pi_{i,i+1})b^2$, is the probability that the two points will lie in different cells and that both cells are black; clearly, since the cells are colored independently and each has probability b of being black, the probability that any two cells will both be black is b^2. We therefore see that $p(B_i, B_{i+1})$ is a function of $\pi_{i,i+1}$ and b only, and it remains to show that $\pi_{i,i+1}$ is the same for all possible pairs of adjacent transect points. Now, the probability that the ith and $(i+1)$th points will lie in the same cell, namely $\pi_{i,i+1}$, is simply the probability that none of the random lines intersects the transect between them. And since the location of these network lines is random it is clear that $\pi_{i,i+1} = \pi$, a constant for all i; that is, every segment of the transect between two adjacent sampling points has the same probability of not being cut by a random line of the network. Thus $p(B_i, B_{i+1})$ does not depend on i and we may put

$$p(B_i, B_{i+1}) \equiv bp_{BB}, \quad \text{say,}$$

where p_{BB} is the *conditional* probability that any point will be a B, given that the preceding point was a B. In other words, the probability that the second point of a pair will be black depends only on whether the first is black and *mutatis mutandis* for the other three ordered pairs of colors, BW, WB, and WW. The transition probabilities for the sequence may therefore be written as a Markov matrix:

$$\mathbf{P} = \begin{matrix} \text{First point} \\ \text{of pair} \end{matrix} \begin{cases} \text{Black} \\ \text{White} \end{cases} \overset{\overbrace{\begin{matrix}\text{Black} & \quad\text{White}\end{matrix}}^{\text{Second point of pair}}}{\begin{pmatrix} p_{BB} & p_{BW} \\ p_{WB} & p_{WW} \end{pmatrix}}. \tag{12.2}$$

For a given mosaic the magnitude of these probabilities will depend only on the distance between the sampling points along the transect: as this distance is decreased, obviously p_{BB} and p_{WW} must increase. The sequence of states along a transect—black patches alternating with white gaps—represents, indeed, a continuous Markov process; instead of making observations of phase at separate equidistant points and treating the observed sequence as a discrete Markov chain, we could measure the lengths cut off on the transect by patches and gaps alternately. These form a realization of a continuous two-state Markov process.

The fact that L-mosaics have the Markov property makes them especially amenable to investigation. It is, conceptually, a straightforward matter to estimate the parameters of any particular mosaic of this form and to derive others of its properties. We pursue this investigation in Section 4, but before doing so we shall describe another kind of random mosaic and compare it with the L-mosaic.

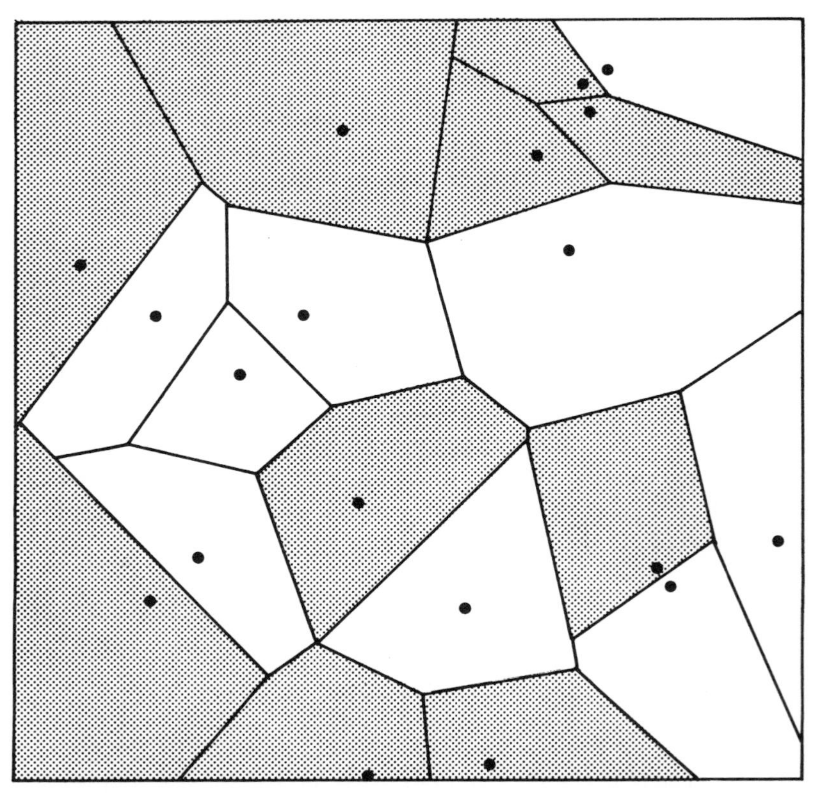

Figure 18. A random-sets mosaic or "S-mosaic."

3. The Random-Sets Mosaic, or S-Mosaic

Matérn (1960) has described in detail the properties of what he calls a "random-sets" mosaic. For brevity we call it an S-mosaic. It is constructed as follows: a pattern of random dots is first drawn in the map area. Then with each dot is associated a cell consisting of all those points in the area that are nearer to that particular dot than to any other. Each segment of a cell boundary is thus the locus of points that are equidistant from the two nearest dots. The cells are then independently colored, as with the L-mosaic, black with probability b and white with probability w, the probabilities being the same for all cells. Figure 18 is an example.

4. The Comparison of L-Mosaics and S-Mosaics

One way of comparing the two kinds of mosaic is to compare their theoretical autocorrelation functions. Unfortunately, for S-mosaics this function can be expressed in elementary form only in the one-dimensional case. Here, therefore, we derive and compare the autocorrelation functions only for one-dimensional forms of the mosaics.

We begin by describing results that are true for both kinds of mosaic of any dimensionality. Suppose we assign arbitrary scores to the two different phases, for example 1 to the black phase (the patches) and 0 to the white phase (the gaps). If the mosaic is now sampled at many ordered pairs of points, the interpoint distance being constant and equal to v, say, a sample of "score-pairs" will be obtained. The expected results can be set out in a 2×2 correlation table as follows:

		Second point: Black (1)	Second point: White (0)	
First point	Black (1)	a_{11} (1)	a_{12} (0)	a_1
	White (0)	a_{21} (0)	a_{22} (0)	a_2
		a_1	a_2	1 .

Here the a_{ij} $(i, j = 1, 2)$ are the expected proportions of the observations that fall into the four classes; they are functions of v, the interpoint distance. The numbers in brackets underneath are the product scores. Corresponding row and column totals obviously have equal expectations from considerations of symmetry and consequently $a_{21} = a_{12}$.

The expected autocorrelation is therefore

$$\rho(v) = \frac{a_{11} - a_1{}^2}{a_1 a_2},$$

and we must now evaluate the right-hand side of this equation as a function of v. Suppose, as before, that the black and white phases are present in proportions b and w; that is, $a_1 = b$ and $a_2 = w$. Denote by $\pi(v)$ the probability that two points a distance v apart will fall in the same cell of the network from which the mosaic was constructed. Then from (12.1), provided $\pi(v)$ is constant everywhere in the mosaic, it is seen that

$$a_{11} = b\pi(v) + b^2[1 - \pi(v)].$$

From similar arguments

$$a_{12} = a_{21} = bw[1 - \pi(v)]$$

and

$$a_{22} = w\pi(v) + w^2[1 - \pi(v)],$$

so that

$$\rho(v) = \pi(v).$$

The requirement that $\pi(v)$ be constant amounts to saying that every cell must have the same expected size, regardless of its location. (The size of a cell is its area if the mosaic is two-dimensional or its length if the mosaic is one-dimensional.) This is not the same as saying that the sizes of contiguous cells are independent of one another, an assertion that is true only for L-mosaics. Indeed, this assertion is what we proved on page 143 and it accounts for the fact that L-mosaics have the Markov property. For an S-mosaic it is easily seen that the sizes of contiguous cells are correlated. However, since the *unconditional* expectation of cell size is the same for all cells, $\pi(v)$ is constant throughout an S-mosaic and the above formulas hold.

The results given so far are quite general. They apply to both L- and S-mosaics, regardless of dimensionality. Thus we have now shown that both mosaics have autocorrelation function $\rho(v) = \pi(v)$; that is to say, the correlation between the phases at two points a given distance apart is equal to the probability that both points are in the same *cell* of the network. In other words, it depends only on the sizes of the network cells and not on b and $w = 1 - b$, the areal proportions of the black and white phases of the mosaic. If the cells are small, the correlation is low and the mosaic may be described as "fine-grained." Conversely, in a "coarse-grained" mosaic the cells are large and the correlation high. This amounts to saying that in a mosaic in which the cells are assigned their colors independently the "grain" of the mosaic depends only on the grain of the cell network. It now remains to determine $\pi(v)$ for one-dimensional mosaics of the two types.

Take L-mosaics first. A one-dimensional L-mosaic consists simply of a row of contiguous nonoverlapping linear cells (i.e., line segments) whose boundaries (which are points) are a realization of a one-dimensional Poisson point process. The lengths of the cells therefore have pdf $f(x) = \lambda e^{-\lambda x}$, where λ is the parameter of the process and is the expected number of points per unit length of line.

Now let a sampling segment of length v (hereafter called simply the segment) be laid at random on the line. The probability that the segment will fall entirely within one cell is $\pi(v)$ and may be determined as follows.

Clearly $\pi(v) = 0$ if the midpoint of the segment falls in a cell of length $x < v$. Suppose it falls in a cell of length $x \geq v$. Then the probability that the whole segment will lie within the cell is easily seen to be $(x - v)/x$. Thus, writing p_x for the probability that the segment's midpoint will fall in a cell of length x, we see that for $x \geq v$,

$$\pi(v) = \frac{x - v}{x} \cdot p_x .$$

It should now be noted that $p_x \neq f(x)$; that is, to select a point at random on the one-dimensional mosaic is *not* equivalent to selecting a cell at random from the population of all cells. Clearly, a point placed at random on the mosaic is more likely to fall in a long cell than a short one and thus the probability that any given cell will contain the segment's midpoint is proportional to the cell's length; that is $p_x = Cx\, f(x)$, where C is a constant of proportionality. Since we must have $\int_0^\infty p_x \, dx = 1$, it follows that

$$C = \frac{1}{\int_0^\infty x\, f(x)\, dx} = \frac{1}{E(x)} = \lambda.$$

Therefore

$$\begin{aligned} p_x &= \frac{x\, f(x)}{E(x)} \\ &= \lambda^2 x e^{-\lambda x}. \end{aligned} \tag{12.3}$$

Then for fixed $x \geq v$

$$\pi(v \mid x) = \frac{x - v}{x} \lambda^2 x e^{-\lambda x}.$$

Integrating over all $x \geq v$ gives

$$\pi(v) = \int_v^\infty \lambda^2(x-v)e^{-\lambda x}\,dx = e^{-\lambda v}.$$

We have now proved that for a one-dimensional L-mosaic the autocorrelation function is

$$\rho(v) = \pi(v) = e^{-\lambda v}.$$

For a two-dimensional L-mosaic $\rho(v)$ is of the same form. This follows because a transect across a two-dimensional L-mosaic is itself a one-dimensional L-mosaic. Both are realizations of a continuous Markov process.

Next we determine $\pi(v)$ [hence $\rho(v)$] for a one-dimensional S-mosaic. A diagram of a linear S-mosaic is shown below. The X's on the line are points generated by a Poisson point process of intensity λ; the dotted lines, which bisect the intervals between the X's, are the cell boundaries. The lengths of

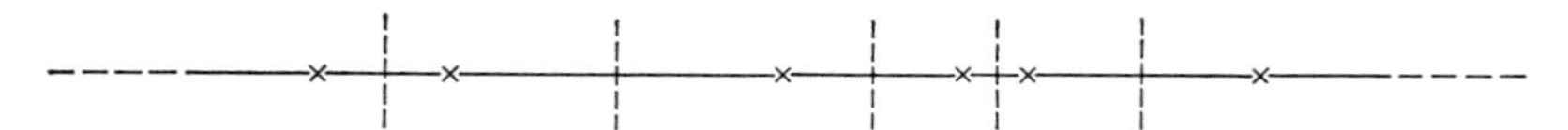

the intervals between the X's have, as before, the pdf $f(x) = \lambda e^{-\lambda x}$. The length of a cell, say y, is therefore the mean of two independent values of x, so y has pdf

$$g(y) = (2\lambda)^2 y e^{-2\lambda y}$$

[see (10.3)] with

$$E(y) = \frac{1}{\lambda}.$$

We now obtain p_y, the probability that the midpoint of a randomly placed sampling segment will fall within a cell of length y. By the same argument that led to (12.3) it is seen that $p_y = yg(y)/E(y)$ so that in this case

$$p_y = 4\lambda^3 y^2 e^{-2\lambda y}.$$

Again, the probability that a sampling segment of length v will fall entirely within a cell of given length y is $(y-v)/y$ for $y > v$, and 0 otherwise. Therefore

$$\begin{aligned}\pi(v) &= \int_v^\infty \frac{y-v}{y} \cdot p_y\,dy \\ &= \int_v^\infty 4\lambda^3(y^2 - vy)e^{-2\lambda y} \cdot dy \\ &= (1+\lambda v)e^{-2\lambda v},\end{aligned}$$

and, since $\rho(v) = \pi(v)$, this is the autocorrelation function for a linear S-mosaic.

Writing the autocorrelation functions for L- and S-mosaics as $\rho_L(v)$ and $\rho_S(v)$, we see that $\rho_L(v) > \rho_S(v)$ for all v and λ. This follows since

$$\begin{aligned}\rho_L(v) - \rho_S(v) &= e^{-\lambda v} - (1 + \lambda v)e^{-2\lambda v} \\ &= e^{-2\lambda v}(e^{\lambda v} - 1 - \lambda v) \\ &= e^{-2\lambda v} \sum_{j=2}^{\infty} \frac{(\lambda v)^j}{j!} > 0.\end{aligned}$$

Therefore, for given λ, the correlation between the phases at a chosen fixed distance apart is greater in a linear L-mosaic than in a linear S-mosaic. Also, given a mosaic of each kind with the same mean cell length, the variance of the cell lengths is twice as great in the L-mosaic as in the S-mosaic; this results from the fact that every cell in an S-mosaic is made up of two adjacent half-cells of an L-mosaic.

It is intuitively clear that in two-dimensional mosaics with equal mean cell areas the autocorrelation will be greater in L-mosaics than in S-mosaics, and, likewise, the cell areas will have greater variance in L-mosaics.

5. The L-Mosaic as a Standard of Randomness

Because of its greater mathematical tractability, the L-mosaic provides a better standard of randomness than the S-mosaic. Although it would be absurd to suppose that the boundaries of the patches in ecological maps resulted from the existence of random lines, actually occurring mosaics may often have patterns that are indistinguishable in many though not all of their properties, from two-phase two-dimensional L-mosaics. Admittedly, we should not expect the boundaries of the patches in a vegetation map, for example, to have sharp angles at the corners and straight edges between, as has a geometrically constructed L-mosaic (see Figure 17), but a mosaic in which the means and variances of the sizes of the patches and gaps are closely approximated by those of an L-mosaic is perfectly possible. If we choose to define such a mosaic as random, we have a useful definition of randomness. A precisely defined concept of randomness is as desirable for mosaic patterns as it is for dot patterns. It provides a standard with which natural mosaics may be compared. Also it should be possible, although this has not yet been done, to devise appropriate ways of measuring the departures from randomness that are exhibited by actual mosaics. Without such a standard, comparisons are difficult. This explains and justifies the preoccupation of ecologists with random patterns in general.

The contrast between dot maps and mosaic maps is this: genuine random dot patterns can certainly occur naturally, although they are uncommon; such a pattern will be found, for instance, whenever a sparse population of small annual plants grows in a homogeneous area from independently dispersed seeds. In contrast, true L-mosaics are undoubtedly nonexistent; it is hard to think of any natural process that would yield a patch and gap pattern with all the geometrical properties of an L-mosaic. This does not, however, detract from the value of an L-mosaic treated as a standard or basis for comparison.

Only one parameter is needed to specify a random dot pattern fully. This is the expected number of dots per unit area or, equivalently, the intensity of the Poisson point process in the plane that generated the pattern. Mosaics, on the other hand, need two parameters before they can be exactly specified. Thus suppose a unit of length has been chosen. In terms of this unit the pair of parameters necessary to specify a particular two-phase L-mosaic may be expressed in two ways:

1. If the mosaic is sampled along a row of equidistant points a unit distance apart, then, as already shown, the sequence of phases encountered (the B's and W's) are a realization of a two-state Markov chain with matrix **P**, as given in (12.2). The pattern of the mosaic is completely specified when any two independent transition probabilities, say p_{BW} and p_{WB}, are given.
2. Any transect across the mosaic will cut through black patches and white gaps alternately. Denote by $E(l_B)$ and $E(l_W)$, the mean lengths along the transect of these patch and gap intercepts. Then, if $E(l_B)$ and $E(l_W)$ are given, the pattern of the mosaic is fully specified.

We now consider how the transition probabilities of **P**, and the linear dimensions of the patches and gaps, are interrelated.

We begin by considering the distribution of the variate l_B, the length, as measured along a transect, of a black patch. Recall that each patch consists of one or more contiguous network cells that chance to have been colored black. The length of any patch is therefore the sum of the lengths of one or more contiguous black network cells. The lengths of these cells are independent. Writing x for the length of a *cell*, x has pdf $f(x) = \lambda e^{-\lambda x}$, where λ is the expected number of network lines cutting across the transect per unit length. We now wish to derive the pdf of l_B, the length of a *patch*.

If a patch consists of j contiguous cells (along the sampling transect), its length is the sum of j independent values of x; so the conditional pdf of l_B for given j is

$$g(l_B \mid j) = \frac{\lambda^j l_B^{j-1} e^{-\lambda l_B}}{(j-1)!}$$

(see page 113). Now, the probability of encountering an uninterrupted sequence, or run, of j black cells along a transect is $b^{j-1}w$. For suppose a black cell has been entered; a run of j black cells will occur if the succeeding $j-1$ cells are also black and then a white cell follows to terminate the run. The probability that this will happen is evidently $b^{j-1}w$.

The unconditional pdf of l_B, when j is allowed to vary, is therefore

$$\begin{aligned} g(l_B) &= \sum_{j=1}^{\infty} g(l_B \mid j) b^{j-1} w \\ &= \lambda w\, e^{-\lambda l_B} \sum_{j=1}^{\infty} \frac{(\lambda b l_B)^{j-1}}{(j-1)!} \\ &= \lambda w\, e^{-\lambda w l_B}, \end{aligned}$$

since $b + w = 1$, or

$$g(l_B) = \lambda_B\, e^{-\lambda_B l_B}, \text{ say,}$$

on putting $\lambda w = \lambda_B$. Thus the pdf of the length of a patch has the same form as that of the length of a single cell. The expected patch length is $E(l_B) = 1/\lambda_B$. Correspondingly, the pdf of l_W, the length of the gaps, is

$$g(l_W) = \lambda_W\, e^{-\lambda_W l_W}$$

where

$$\lambda_W = b\lambda \quad \text{and} \quad E(l_W) = \frac{1}{\lambda_W}.$$

Further,

$$\lambda_B + \lambda_W = \lambda(b + w) = \lambda,$$

the density of the network lines.

We have now shown that both the patch lengths and the gap lengths are exponentially distributed, with means $1/\lambda_B$ and $1/\lambda_W$, respectively. An alternative way of stating the same result is to say that the alternating patch lengths and gap lengths constitute a realization of a continuous two-state Markov process. The parameters λ_B and λ_W are known as the transition rates of the process, and the matrix **R**, defined as

$$\mathbf{R} = \begin{pmatrix} -\lambda_B & \lambda_B \\ \lambda_W & -\lambda_W \end{pmatrix}$$

is the matrix of transition rates.

Now it may be shown (e.g., see Howard, 1960) that

$$\mathbf{P} = \mathbf{I} + \mathbf{R} + \frac{\mathbf{R}^2}{2!} + \frac{\mathbf{R}^3}{3!} + \cdots. \tag{12.3}$$

Thus the transition *probabilities* (the elements of **P**) may be expressed in terms of the transition *rates*, and vice versa, by equating corresponding elements on both sides of (12.3). Writing, as before, $\lambda_B + \lambda_W = \lambda$, we see that $\mathbf{R}^2 = -\lambda\mathbf{R}$ and consequently that $\mathbf{R}^{j+1} = (-1)^j\lambda^j\mathbf{R}$.

Therefore (12.3) may be written

$$\mathbf{P} = \mathbf{I} + \mathbf{R} - \frac{\lambda}{2!}\mathbf{R} + \frac{\lambda^2}{3!}\mathbf{R} - \frac{\lambda^3}{4!}\mathbf{R} + \cdots$$

$$= \mathbf{I} - \frac{(e^{-\lambda} - 1)}{\lambda}\mathbf{R}.$$

It now follows that

$$p_{BW} = \frac{1 - e^{-\lambda}}{\lambda} \cdot \lambda_B \quad \text{and} \quad p_{WB} = \frac{1 - e^{-\lambda}}{\lambda} \cdot \lambda_W .$$

Also, $p_{BW} + p_{WB} = 1 - e^{-\lambda}$, hence $\lambda = -\ln(1 - p_{BW} - p_{WB})$.

Therefore

$$\lambda_B = \frac{-p_{BW}}{p_{BW} + p_{WB}} \ln(1 - p_{BW} - p_{WB})$$

and

$$\lambda_W = \frac{-p_{WB}}{p_{BW} + p_{WB}} \ln(1 - p_{BW} - p_{WB}).$$

It will be recalled that $\lambda = \lambda_B + \lambda_W$ is the density of the network lines from which the mosaic is constructed. It is a measure of the grain of the mosaic (see page 147) and is independent of the areal proportions of the patch and gap (or black and white) phases. When λ is large, the sizes of the patches or gaps (or both) are small and the mosaic is fine-grained; when λ is small, the patches and/or gaps are large and the mosaic is coarse-grained.

6. Practical Problems of Sampling Mosaics

To test whether a natural mosaic can be regarded as random we must either (a) obtain a sample of values of l_B and l_W and judge whether their distributions are fitted by the pdf's $g(l_B)$ and $g(l_W)$; or (b) sample the mosaic along a row of equidistant points and judge whether the observed sequence of phases forms a realization of a Markov chain. Although method (a) is conceptually more straightforward, method (b) is preferable in practice because natural mosaics rarely have clear-cut boundaries between the phases. When a vegetation mosaic is sampled in the field, it is usually difficult to decide exactly where a patch ends and the succeeding gap begins, so that measurements of the sample values of l_B and l_W are hard to make; even if a map of the mosaic has been drawn, making measurement easy, the results are probably unreliable and the seeming accuracy illusory. It therefore seems better always to use method (b). Methods of estimating the transition probabilities and of fitting the Markov hypothesis to the data are given in Pielou (1964).

When one is sampling natural vegetation on the ground, difficulties often arise with plant species that form very open clumps; the above-ground shoots of a vegetatively reproducing plant may be widely spaced, and it is unreasonable to treat the spaces among them as part of the gap phase. The best way to deal with this problem is to sample a mosaic along a row of very small circular quadrats instead of at true points; then, when any plant part falls within a circular quadrat, the center of the quadrat is treated as being within a patch. The choice of radius for the quadrats is necessarily arbitrary; their use will always lead to overestimation of the area of the patch phase, and the larger the quadrats, the greater the overestimation. To use circular quadrats instead of points is, in effect, to blur the pattern; it is as though one were examining the mosaic with an optical instrument of low resolving power. However, this is not necessarily a disadvantage. By deliberately using low resolution, we can prevent the minor details of a pattern from obscuring its major features.

7. Non-Random Mosaics and Anisotropic Mosaics

For a mosaic to be random in the sense we have defined it is necessary for both the patch lengths and the gap lengths to be exponentially distributed. Therefore a mosaic may be random in one of its phases and nonrandom in the other; for example, the gap lengths could have a variance either less than or greater than expected even though the patch lengths had an exponential distribution. The patches could then be said to be regularly spaced, or aggregated, in relation to the gaps. The reverse is also possible; the patch lengths may vary either to a lesser or a greater extent than expected, regardless of the distribution of the gap lengths. Any mosaic can, in fact, be thought of as having two patterns: that of the patches relative to the gaps and that of the gaps relative to the patches. One cannot say of a mosaic, as one can of a dot pattern, merely that its pattern is regular, random or aggregated. Both the patches and the gaps have definable patterns and nine combinations can be envisaged.

The possibility that a mosaic may not be isotropic should also be borne in mind. If the mean lengths of the patches or gaps (or both) are longer in one direction than in another the mosaic is anisotropic. There are a number of possible causes of anisotropy in vegetation mosaics; for instance, elongated ridges and hollows, an anisotropic soil pattern, or the effects of the prevailing wind in an exposed situation may all produce vegetation patches that are not isodiametric. The data yielded by transect sampling will then vary with the direction of the transect.

8. Estimated Maps

It is obviously conceptually possible to map any natural mosaic pattern. If the thing to be mapped is visible and shows up clearly in a vertical photograph, a map can easily be made by tracing the photograph, but if one wishes to map such invisible features as height of the water table, depth of litter, or soil texture, for example, the usual way to proceed is to sample the ground at a number of points, mark the observed result at corresponding points on the map sheet, and attempt to reconstitute the pattern from the points. Switzer (1967) has investigated the accuracy of such estimated patterns and has explored the effect on accuracy of the different arrangements and spacings of the sample points.

We shall suppose that a two-colored map is to be made or, in other words, that two phases are to be distinguished. These phases may differ in the presence or absence of an attribute or, if the variate is continuous, in being high-valued or low-valued. The accuracy of a two-color estimated map may be measured by a loss function, L, defined as the proportion of the total area misclassified on the map. Misclassified areas consist of those parts shown black that should be white and vice versa. Switzer has derived the expected

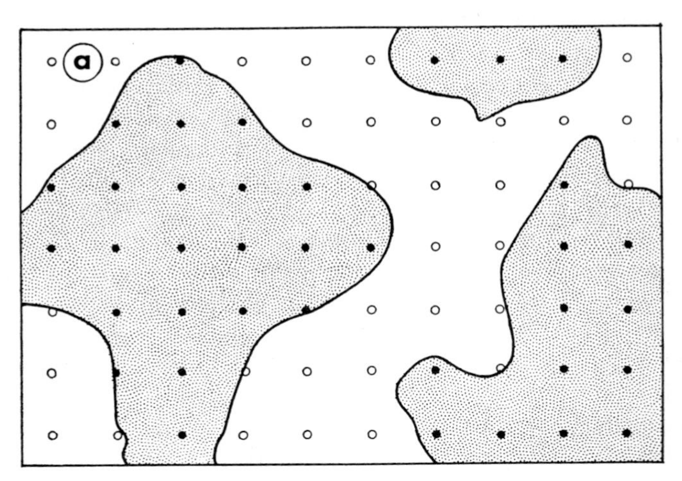

Figure 19*a*. A two-phase mosaic pattern sampled at a square lattice of points. These points are shown as black or white circles according as they occur in the black or white phase.

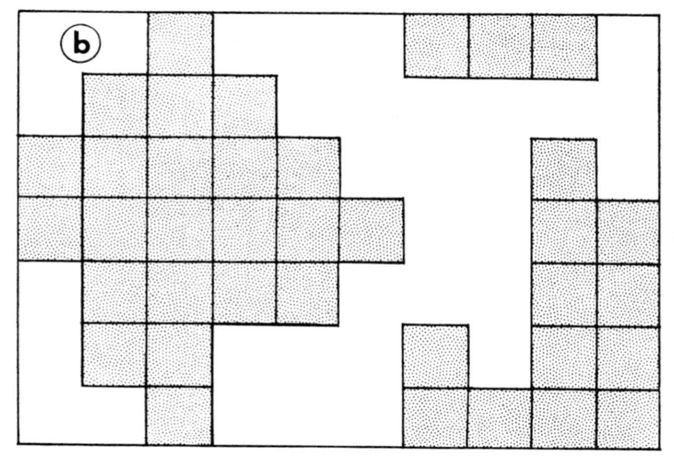

Figure 19*b*. The estimated map.

loss function EL for different kinds of mosaics mapped by various sampling methods, but here it will be possible to quote only two of his results without proof.

1. Suppose the area to be mapped is rectangular and can be subdivided into n squares each of area $1/n$. The area is sampled at the center of each square and the corresponding square on the map is colored black or white in accordance with the phase observed at the sampling point (see Figure 19). Then, if the pattern is an L-mosaic and if the proportions of the black and white phases on the ground are b and $w = 1 - b$,

$$\frac{EL}{2bw} = 1 - 8\int_0^{\frac{1}{2}}\int_0^y \exp\left[-\lambda\left(\frac{x^2+y^2}{n}\right)^{1/2}\right] dx\, dy.$$

2. Again assume that the pattern to be mapped is an L-mosaic but suppose now that it can be sampled continuously (rather than at a succession of points) along parallel transects a distance d apart. This is called line sampling with spacing d. The lines can be drawn on a map and marked to show the intercepts of the black and white phases as they occurred on the ground. Each line is then expanded to form a strip of width d with the sampling transect along its center (see Figure 20). For the estimated map that results it can be shown that

$$\frac{EL}{2bw} = 1 - \frac{2(1 - e^{-\lambda d/2})}{\lambda d}.$$

Further details and other results have been obtained by Switzer and will be found in his paper. Their potential usefulness to ecologists is obvious.

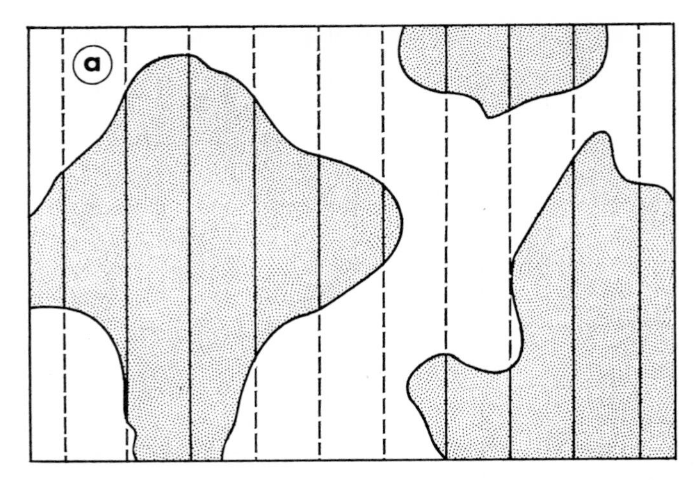

Figure 20*a*. A two-phase mosaic pattern sampled along the marked transect lines.

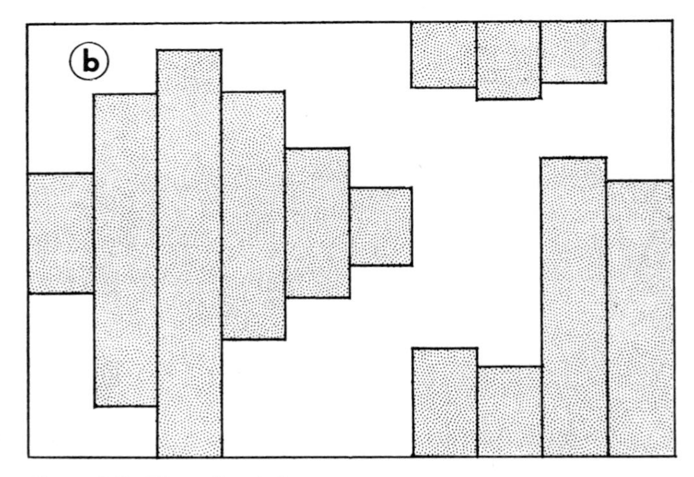

Figure 20*b*. The estimated map.

III

Spatial Relations of Two or More Species

13

Association Between Pairs of Species I: Individuals in Discrete Habitable Units

1. Introduction

The spatial pattern exhibited by a single species within a limited area is often worth examining for its own sake. The factors controlling and determining pattern, however, are likely to affect many species rather than just one, and much may be learned by investigating the way in which species are associated with one another. If two co-occurring species are affected by the same environmental factors, or if they have some effect, either favorable or unfavorable, on each other, their patterns will not be independent; the species will be associated, either positively or negatively. Association or the lack of it among pairs and groups of species is therefore of obvious ecological interest. As in the study of pattern in one-species populations, it is desirable to consider separately those species that occupy discrete habitable units (e.g., pest insects in fruits, ectoparasites on rats) and those that may occur anywhere throughout an extended space or continuum (e.g., plankton organisms in a volume of water, plants in a meadow).

In this chapter we confine attention to organisms in discrete units and begin by discussing the association of a single pair of species. The far more difficult problem of investigating association in a group of more than two species is mentioned briefly in Sections 6 and 7.

2. Testing the Association Between Two Species

The standard method of testing for association between two species is as follows. Assume that we are examining a sample of N discrete units collected at random from a large population of possible units. Let the two species being studied by labeled species A and species B; for each unit note whether it contains species A alone, species B alone, both species, or neither species.

The quantity of each species in each unit is disregarded; we record only presences and absences. The observed frequencies can then be set out in the form of a 2×2 table:

		Species B		
		Present	Absent	
Species A	Present	a	b	$m = a + b$
	Absent	c	d	$n = c + d$
		$r = a + c$	$s = b + d$	$N = m + n = r + s.$

The usual approach of ecologists to 2×2 tables is to carry out a χ^2-test and let it go at that. This is a slovenly approach, and it is important to realize that there are two entirely different questions we could ask on being confronted with a table such as this (see Pearson, 1947).

Question 1. Among the N units examined do species A and species B occur independently of each other?
Question 2. In the population as a whole are the two species independent of each other?

Assume first that Question 1 is being asked. We *know* that m of the N units contain species A and that r of them contain species B; the table's marginal totals are therefore fixed. The question thus becomes: for these given marginal totals what are the probabilities of the various possible sets of cell frequencies (or partitions of N)? Is the particular set of cell frequencies we have observed consistent with the hypothesis of independence? For given N, m, and r, the conditional probability that a of the units will contain both species is

$$\Pr(a \mid N, m, r) = \frac{m!\,n!\,r!\,s!}{a!\,b!\,c!\,d!\,N!};$$

that is, a has a hypergeometric distribution. To see this note that the number of ways of choosing m units out of N to contain species A is $\binom{N}{m}$; similarly, the number of ways of choosing r units to contain species B is $\binom{N}{r}$. Thus the number of arrangements that would give rise to the observed marginal totals is $\binom{N}{m}\binom{N}{r}$.

The number of different ways of partitioning N to produce the observed cell frequencies a, b, c, and d is $N!/(a!\,b!\,c!\,d!)$. Therefore

$$\Pr(a \mid N, m, r) = \frac{N!/(a!\,b!\,c!\,d!)}{\binom{N}{m}\binom{N}{r}} = \frac{m!\,n!\,r!\,s!}{a!\,b!\,c!\,d!\,N!}. \tag{13.1}$$

In this way we may calculate the probabilities for all the different sets of cell frequencies that give rise to the observed marginal totals.

We resume discussion of this case after considering how the second case (Question 2) differs from it. In asking Question 2, it is no longer assumed that the marginal totals are fixed. When a sample of N units is taken at random from a large population of units, not only are the cell frequencies free to vary but so also are their pairwise sums, the marginal totals. To determine the probability of obtaining any particular 2×2 table we must argue as follows.

Let $p(A)$ denote the probability that a unit will contain species A and $p(\bar{A}) = 1 - p(A)$, the probability that a unit will lack species A. The probabilities $p(B)$ and $p(\bar{B}) = 1 - p(B)$ are defined likewise for species B. Any unit must belong to one of four classes, AB, $A\bar{B}$, $\bar{A}B$, or $\overline{AB}$, and on the null hypothesis of independence of the species we must have

$$p(AB) = p(A) \cdot p(B), \qquad p(A\bar{B}) = p(A) \cdot p(\bar{B}),$$

$$p(\bar{A}B) = p(\bar{A}) \cdot p(B), \qquad p(\overline{AB}) = p(\bar{A}) \cdot p(\bar{B}).$$

The probability $\Pr(a, b, c, d)$ of obtaining the observed cell frequencies a, b, c, and d in a sample of N units is then a term from a multinomial distribution; it is given by the coefficient of $z_1^a\, z_2^b\, z_3^c\, z_4^d$ in the expansion of the probability generating function

$$[p(AB)z_1 + p(A\bar{B})z_2 + p(\bar{A}B)z_3 + p(\overline{AB})z_4]^N;$$

that is

$$\begin{aligned}
\Pr(a, b, c, d) &= \frac{N!}{a!\,b!\,c!\,d!}[p(AB)]^a[p(A\bar{B})]^b[p(\bar{A}B)]^c[p(\overline{AB})]^d \\
&= \frac{N!}{a!\,b!\,c!\,d!}[p(A)]^{a+b}[p(B)]^{a+c}[p(\bar{A})]^{c+d}[p(\bar{B})]^{b+d} \\
&= \frac{N!}{m!\,n!}[p(A)]^m[1 - p(A)]^n \\
&\qquad \cdot \frac{N!}{r!\,s!}[p(B)]^r[1 - p(B)]^s \cdot \frac{m!\,n!\,r!\,s!}{a!\,b!\,c!\,d!\,N!} \\
&\equiv b(m \mid p(A), N) \times b(r \mid p(B), N) \times \Pr(a \mid N, m, r). \qquad (13.2)
\end{aligned}$$

Here the binomial term $b(m \mid p(A), N)$ denotes the probability that in N trials an event whose probability is $p(A)$ will occur m times; likewise $b(r \mid p(B), N)$ is the probability of r occurrences in N trials of an event with probability $p(B)$, and $\Pr(a \mid N, m, r)$ is the conditional probability we found in answering Question 1, namely, the probability of observing the partition of N into the parts a, b, c, and d, given that $a + b = m$ and $a + c = r$.

The probability of obtaining an observed 2×2 table thus depends on whether the table is assumed to have preassigned marginal totals, in which case it represents what Barnard (1947) has called a doubly restricted double dichotomy; this is the assumption made when Question 1 is asked. Or whether the marginal totals as well as the cell frequencies are treated as random variates, giving a table that is an unrestricted double dichotomy; this is the assumption made when Question 2 is asked.

3. Testing the Association in a Sample (Question 1)

If the empirical table is treated as a doubly restricted double dichotomy, we may calculate $\Pr(a \mid N, m, r)$ for all the possible values of a that could arise, subject to the restriction that the marginal totals are fixed. Denote the minimum and maximum possible values of a by $a(\text{min})$ and $a(\text{max})$.

Now suppose that the observed frequency a of the event AB (i.e., the observed number of joint occurrences of species A and B) is greater than its expectation $E(a)$; that is, $E(a) < a \leq a(\text{max})$. This leads us to believe that there may be significant positive association between the species. Then the probability of observing a deviation from expectation as great as or greater than $a - E(a)$, and in the same direction, is

$$P_{\text{upper}} = \sum_{i=a}^{a(\text{max})} \Pr(i \mid N, m, r).$$

Thus P_{upper} is the appropriate probability for a one-tail test for positive association. It is the probability, on the null hypothesis of independence, of obtaining evidence for positive association as strong as or stronger than that observed.

Likewise, if $a(\text{min}) \leq a < E(a)$, so that the data suggest negative association, the probability required for a one-tail test is

$$P_{\text{lower}} = \sum_{i=a(\text{min})}^{a} \Pr(i \mid N, m, r).$$

To do a two-tail test we must sum the probabilities of obtaining a deviation as great as or greater than that observed in either direction. Thus, if the deviation of the observed a from expectation, $|E(a) - a|$, is x, the probability for the two-tail test is

$$\left\{ \sum_{i=a(\text{min})}^{E(a)-x} + \sum_{i=E(a)+x}^{a(\text{max})} \right\} \Pr(i \mid N, m, r).$$

This *exact* test is easily done with the help of tables such as those of Finney et al. (1963) and Bennett and Horst (1966), provided both column totals (or both row totals) are ≤ 50. Outside the range of these tables we may make use of the fact that the distribution of a tends to normality. As already remarked, a is a hypergeometric variate. Its mean and variance are

$$E(a) = \frac{rm}{N} \quad \text{and} \quad \text{var}(a) = \frac{mnrs}{N^2(N-1)}.$$

So we may treat

$$X = \frac{a - E(a)}{\sqrt{\text{var}(a)}}$$

as a standardized normal variate and refer to Normal tables to judge significance. Either a one-tail or a two-tail test may be done.

For large N it is permissible to substitute N for $N - 1$ in the denominator of var(a). Then X becomes

$$X = \frac{\sqrt{N}(ad - bc)}{\sqrt{mnrs}} \quad \text{and} \quad X^2 = \frac{N(ad - bc)^2}{mnrs}. \tag{13.3}$$

We see that X^2, being the square of a standardized normal variate, has the χ^2-distribution with one degree of freedom.

Since the continuous χ^2-distribution is being used to approximate a discrete distribution, it is desirable to make a continuity correction. In calculating X^2 this is done by subtracting $\frac{1}{2}$ from the two observed frequencies that exceed expectation and adding $\frac{1}{2}$ to the two frequencies that fall short of expectation. Then

$$X^2(\text{corrected}) = \frac{[|ad - bc| - N/2]^2 N}{mnrs}.$$

This ensures a closer approximation of the χ^2-integral to the sum of the tail terms of the true, discrete distribution of X^2.

It will be seen that the expression here denoted by X^2 is the one often described as χ^2. However, for a function of the observed cell frequencies (in other words, a sample statistic) it is preferable to use the noncommital symbol X^2; the symbol χ^2 should be reserved for the theoretical variate with the χ^2-distribution (see Cochran, 1954).

The foregoing arguments explain why we may use the χ^2-test as an approximation to the exact test appropriate to a doubly restricted double dichotomy. However, ecologists who use the test should never lose sight of the fact that a χ^2-test is automatically two-tailed. So if we sometimes use the χ^2-test and at other times (because of low observed frequencies) the exact test, the two-tail form of the exact test should be used. Otherwise the results are not comparable.

4. Testing the Association in a Population of Units (Question 2)

The test just described, in either its exact or approximate version, is strictly applicable only to a doubly restricted double dichotomy; it answers Question 1. Now suppose we require an answer to Question 2; that is, we wish to know whether the data yielded by a *sample* could have come from a *population* in which the two species are independent. The desired probability is a sum of terms of the form of $\Pr(a, b, c, d)$ in (13.2), in which hypergeometric probabilities are weighted with binomial probabilities. The χ^2-test makes no allowance for the binomial terms and although it is the best test (for a proof of this, see Kendall and Stuart, 1967), the calculated tail probabilities are greater than their true values. This may lead to acceptance of the null hypothesis of independence when it should be rejected—a type II error. At the same time the risk of asserting that there is true association when there is not—a type I error—is reduced.

For sufficiently large samples the error introduced is usually negligible. However, there are two reasons for stressing the conceptual difference between the two situations. In the first place ecologists are often tempted to use the exact test when some of the cell frequencies yielded by a sample are very low. This is satisfactory when one is inquiring whether the species are independent within the particular sample of units being examined. But it is futile to use the exact test when an answer to Question 2 is sought; the exact answer to one question may be very inexact as an answer to a different question.

In the second place certain methods of classifying vegetation (to be described in Chapter 19) require repeated applications of the ordinary χ^2-test to 2×2 tables, often with quite low values of N. A risk of error that can reasonably be ignored when only one or a few tests are being done may then become appreciable.

5. Measurements of Association

Besides testing a 2×2 table to judge whether the null hypothesis of independence should be accepted or rejected, we may also wish to measure the strength of the association between two species. There are many ways of measuring association, some devised by statisticians for use with any 2×2 table; and others by ecologists in search of a coefficient particularly suited to measuring ecological association. Rather than catalog a great many of them, it will be more useful to consider a few in detail. For clarity, we shall speak only of positive association. The modifications to be made in the arguments when the species are negatively associated are obvious.

Two desirable properties of a coefficient to measure association are: (i) that it should be zero when the observed cell frequencies are equal to their expected values; and (ii) that it should range from -1 to $+1$, taking the value -1 when negative association is as great as possible and $+1$ when positive association is as great as possible. The second of these properties is ambiguous: what is meant by "as great as possible"?

The phrase can have two meanings. Suppose species B occurs in more of the units than does species A. Then positive association would be as great as possible if A was never found in the absence of B, though there would perforce be some units in which B was found without A. This degree of association can be called *complete* (see Kendall and Stuart, 1967). However, we might choose to assert that the association was "as great as possible" only when neither species ever occurred without the other. This is called *absolute* association. In terms of the cell frequencies in the 2×2 table complete positive association requires that either b or c (not necessarily both) be zero. For absolute association we must have both $b = 0$ and $c = 0$; then $m = r = a$ and $n = s = d$. Depending on whether we want the coefficient of association to be $+1$ when the association is complete or absolute, we can use the coefficients Q or V defined as follows (Yule, 1912):

$$Q = \frac{ad - bc}{ad + bc};$$

then $Q = 1$ when either $b = 0$ or $c = 0$.

Or
$$V = \frac{ad - bc}{+(mnrs)^{1/2}}.$$

Then $V = \pm 1$ only if $mnrs - (ad - bc)^2 = 0$, but since

$$\begin{aligned} mnrs - (ad - bc)^2 = 4abcd &+ a^2(bc + bd + cd) \\ &+ b^2(ac + ad + cd) + c^2(ab + ad + bd) \\ &+ d^2(ab + ac + bc) \end{aligned}$$

the expression on the right vanishes only if two of the cell frequencies are zero. We can exclude from consideration cases in which the two zeros occur in the same row or the same column, for there would be nothing to test. This leaves either $b = 0$ and $c = 0$, for which $V = +1$, or $a = 0$ and $d = 0$, for which $V = -1$.

Both coefficients are zero when the association is nil, that is, when observed and expected frequencies are equal, for then $a - E(a) = (ad - bc)/N = 0$.

The sampling variance of Q is

$$\text{var}(Q) = \frac{(1 - Q^2)^2}{4} \left\{ \frac{1}{a} + \frac{1}{b} + \frac{1}{c} + \frac{1}{d} \right\}.$$

The derivation is given in Kendall and Stuart (1967). We shall not consider Q further here. The fact that its use precludes any distinction between complete and absolute association makes it unsuitable as a measure of ecological association. To see this consider the two tables

		Species B		
		+	−	
Species A	+	80	80	160
	−	0	15	15
		80	95	175,

		Species B		
		+	−	
Species A	+	80	0	80
	−	0	15	15
		80	15	95.

They are strikingly different. In the population tabulated on the left, although all the 80 units containing B also contain A, there are in addition 80 units with only A in them. By contrast in the population on the right neither species occurs without the other. For both tables $Q = 1$, whereas $V = 1$ only for the right-hand table; for the left-hand table $V = 0.281$. Any ecologist would assert that the association shown by the table on the right was by far the greater and V is therefore preferable to Q in ecological contexts.

Two other points to mention about V are the following: (a) $V^2 = X^2/N$, where X^2 is the test statistic defined in (13.3); V^2 is known as the mean-square contingency of the 2×2 table. (b) V is a correlation coefficient. Assign to each unit a pair of values (x, y) with

$$x = \begin{cases} 1 \text{ when species } A \text{ is present,} \\ 0 \text{ when species } A \text{ is absent,} \end{cases} \qquad y = \begin{cases} 1 \text{ when species } B \text{ is present,} \\ 0 \text{ when species } B \text{ is absent.} \end{cases}$$

Then

$$\text{cov}(x, y) = \frac{a}{N} - \frac{mr}{N^2} = \frac{ad - bc}{N^2},$$

$$\text{var}(x) = \frac{mn}{N^2} \quad \text{and} \quad \text{var}(y) = \frac{rs}{N^2},$$

and therefore

$$V = \frac{ad - bc}{\sqrt{mnrs}} = \frac{\text{cov}(x, y)}{[\text{var}(x)\text{var}(y)]^{1/2}}.$$

In other words, V is the correlation coefficient between x and y. An estimate of its sampling variance, derived by Yule (1912), is given by

$$\text{var}(V) = V^2 \left\{ -\frac{4}{N} + \frac{ad(a + d) + bc(b + c)}{(ad - bc)^2} - \frac{3}{4}\left[\frac{(m - n)^2}{Nmn} + \frac{(r - s)^2}{Nrs}\right] + \frac{(ad - bc)(m - n)(r - s)}{2Nmnrs} \right\}.$$

It should be noticed that although V is a correlation coefficient we cannot use the usual formula for the sampling variance of a correlation coefficient, r, namely, $\text{var}(r) = (1 - r^2)^2/N$, because this formula is appropriate only when the parent distribution of the x's and y's is bivariate normal. In the present case each variate is discrete and can take only the values 0 and 1.

An example of a coefficient of association proposed especially for ecological use is one suggested by Cole (1949) and used by him to measure the association between two species of mites parasitic on rats trapped in food-processing establishments. He stipulated that a desirable coefficient for measuring the degree of association of a pair of species should have the following properties:

1. It should be zero when the observed frequencies are equal to those expected on the null hypothesis of independence; that is, when $a = E(a)$.
2. It should be $+1$ (or -1) when $a - E(a)$ has its maximum possible positive (or negative) value compatible with the observed marginal totals.
3. The coefficient should vary linearly with a.

Condition 3 amounts to requiring (see Figure 21) that the graph of the coefficient C plotted against a should be a straight line going through the

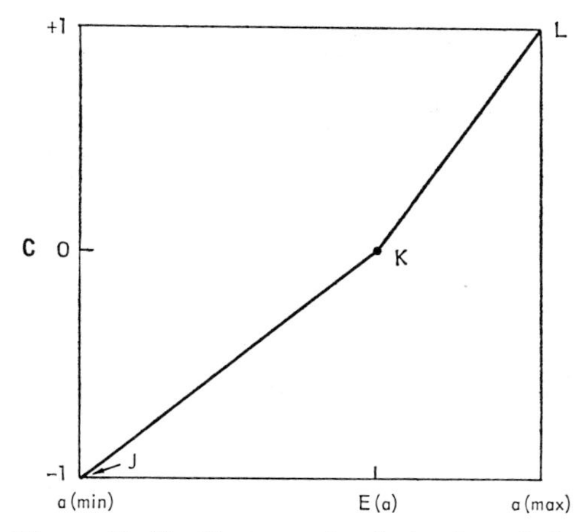

Figure 21. To illustrate the derivation of the different formulas for Cole's coefficient of interspecific association.

points $[a(\text{min}), -1]$, $[E(a], 0]$ and $[a(\text{max}), +1]$, where $a(\text{min})$ and $a(\text{max})$ are the smallest and largest values that a could have. Obviously this requirement cannot be met unless the three points are colinear. This will be so only when $E(a) = \frac{1}{2}[a(\text{min}) + a(\text{max})]$. Cole therefore proposed different formulas for

C for use in different circumstances. We now require that positive values of C shall lie on the line joining $[E(a), 0]$ and $[a(\max), +1]$ and that negative values of C shall lie on the line joining $[a(\min), -1]$ and $[E(a), 0]$.

Suppose first that the association is positive or that $ad > bc$. Then C must fall on the line KL in the figure, and

$$C = \frac{a - E(a)}{a(\max) - E(a)}.$$

Let the species be so labeled that A is the less frequent species or $m \leq r$. Then

$$a(\max) = m \quad \text{and} \quad C = \frac{a - mr/N}{m - mr/N} = \frac{ad - bc}{ms}.$$

Next suppose that there is negative association or that $ad < bc$. Then C must fall on the line JK and

$$C = \frac{a - E(a)}{E(a) - a(\min)}.$$

The value of $a(\min)$ depends on whether $a \leq d$ or $a > d$. If $a \leq d$, $a(\min) = 0$. Writing C_1 for the coefficient in this case,

$$C_1 = \frac{a - mr/N}{mr/N} = \frac{ad - bc}{mr}.$$

If $a > d$, $a(\min) = a - d$. Writing C_2 for the coefficient,

$$C_2 = \frac{a - mr/N}{mr/N - (a - d)} = \frac{ad - bc}{ns}.$$

Cole also obtained the sampling variances of the three versions of his coefficient.

Property (2) of the coefficient (see page 167) requires that

$$C = \begin{cases} +1 \text{ when } a = a(\max), \\ -1 \text{ when } a = a(\min). \end{cases}$$

Therefore, like Yule's Q, it suffers from the defect that no distinction is made between complete and absolute association.

6. Association Among k Species

Up to this point we have discussed association between two species. Now suppose we wish to explore the joint occurrences of several, say k, species. Suppose that for each of a sample of N units it has been noted whether it contains species i $(i = 1, 2, \ldots, k)$. Each unit then belongs to one of 2^k

distinguishably different classes and there are 2^k comparisons to make between observed and expected frequencies. Obviously this is impracticable as soon as k exceeds 3 or 4. However, it may be worthwhile to enquire whether the group of k species *as a whole* is associated. Limiting the discussion to positive association only, for the sake of clarity, we could say that a group of k species was positively associated as a whole if an unexpectedly large number of units contained representatives of all of them. To test whether this were so or not, we should need to compare observed and expected frequencies only in the class defined by joint occurrence of all the species.

Another comparison that could be made is between the observed and expected number of "empty" units, that is, those containing none of the k species. If the observed number exceeds expectation, we can say that the species-occurrences are restricted to unexpectedly few of the units or equivalently that there must be some positive associations and these are masking the effect of negative associations, if any. Notice that an excess of units containing all the species on the one hand and an excess of empty units on the other do not lead to identical conclusions. Only in the first case can we say that *all* the species are positively associated with one another. The second case merely leads one to infer that positive associations preponderate over negative ones and the excess of empty units is the net result of several effects.

In testing for association among several species, it will often be found that at least some of the species occur in only a few of the units. This makes it desirable to do a test that treats the observed frequencies of occurrence of the species as fixed rather than as random variates; these frequencies are the marginal totals of a 2^k table. In other words, we test the significance of the association within the sample actually examined (Question 1) without attempting to infer whether the conclusion holds for the parent population from which the sample was drawn.

To judge whether the number of empty units significantly exceeds expectation the following test may be done. (Obvious modifications yield a test of whether the observed number of units containing all the species exceeds expectation.)

Let there be N units and k species and let n_i of the units contain species i ($i = 1, 2, \ldots, k$). The probability that a unit chosen at random lacks species i is $(N - n_i)/N$. Denoting the number of empty units by m, we see that its expectation is

$$E(m) = N \prod_{i=1}^{k} \left(\frac{N - n_i}{N}\right).$$

Provided N is large and k small, the distribution of m is approximately normal and, as shown by Barton and David (1959), it has variance:

$$\text{var}(m) = E(m) - [E(m)]^2 + N(N-1) \prod_{i=1}^{k} \frac{(N-n_i)(N-n_i-1)}{N(N-1)}.$$

Subtracting $\frac{1}{2}$ from the observed deviation as a continuity correction, we may treat $[m - E(m) - \frac{1}{2}]/\sqrt{\text{var}(m)}$ as a standardized normal variate and judge significance from tables of the normal distribution.

If N is small or k large, we can no longer treat m's distribution as approximately normal. An exact, though laborious, test may still be done by summing the probabilities of the different values of m compatible with the observed n_i. The required probabilities are derived in Barton and David (1959), and Pielou and Pielou (1967) give an example of the application of the test. It is an extension to a 2^k table of the exact test for a 2×2 table described on page 162.

7. Segregative and Non-Segregative Association

Suppose two species have been found to be positively associated and it is suspected that the only cause is that some of the units in the sample were intrinsically unsuitable for either of the species. If this were so, the number of empty units (the frequency d in the 2×2 table) would necessarily be improbably large and would lead inevitably to positive association. Some ecologists would argue that such association was apparent rather than real and that the high value of d was, in a sense, artificial. This conclusion seems unwarranted (see page 176), but, granting its reasonableness for the moment, we may now enquire whether the species are independently distributed among the units that *can* contain them. When only two species are considered, it is impossible to make a judgment merely on the basis of statistical tests, for we need only assume that the d empty units consist of $d - b/ca$ that are "unoccupiable" and bc/a that, though "occupiable," are empty as the result of chance. Substitution of bc/a for the observed d then immediately yields a 2×2 table in which observed and expected frequencies are identical. (Inconsequential adjustments are necessary, of course, if bc/a is not a whole number.)

When more than two species are observed, however, it does become legitimate to ask whether the species are independent of one another within the occupiable units only. This follows, since, given a 2^k table in which the observed number of empty units exceeds expectation, it may or may not be possible to find a number which, if substituted for the observed number of empties, will yield a table not significantly different from that expected on the null hypothesis of independence of the species.

When such a number can be found, there is no reason to reject the hypothesis that the species are independent, since the association could be ascribed simply to the presence of some unoccupiable units. This type of association may be called *nonsegregative*. Conversely, the association can be called *segregative* if no such number exists; that is, if the discrepancy between observed and expected frequencies cannot be attributed merely to the presence of an unknown number of unoccupiable units. The occurrence of segregative association implies either that the various species are responding differently to differences among the units or that the species are affecting one another in various ways. Some ecologists would regard only segregative association as "true" and nonsegregative as "spurious."

A test for judging whether the association among several species is segregative or nonsegregative has been described by Pielou and Pielou (1968). The test is designed for use with species of infrequent occurrence and entails the comparison of the observed and expected numbers of different classes of unit encountered when the units are classified according to the species they contain. The number of units in each class is disregarded.

14

Association Between Pairs of Species II: Individuals in a Continuum

1. Introduction

Studies of the association between species that occupy discrete units normally take no account of the spatial arrangement of the units. The mathematical methods for judging the association between, say, a pair of parasite species infesting a population of mammals are formally identical with those for testing the association between, for example, the eye colors of parents and children. The spatial arrangement of the sample units is disregarded except in so far as the population studied is defined in terms of the geographical area it occupies.

To sample individuals that are scattered through a continuum necessitates taking arbitrarily delimited bits of the continuum as sample units; some of the difficulties that arise were mentioned in Chapter 9, which dealt with the spatial patterns of single species. Continuum sampling is most often done by plant ecologists who use quadrats to study the patterns and interspecies relations of the plants growing on a tract of ground. So, in all that follows we shall, for simplicity, speak of the sample units as quadrats. In testing for association between two plant species, it is customary to treat each quadrat as if it were a discrete sample unit and to use the same methods as those described in Chapter 13.

Most plant ecologists (e.g., Greig-Smith, 1964) are aware of the problems that may arise from treating an arbitrary quadrat as though it were a discrete natural entity, but many authors seem to confound two wholly different sources of difficulty which ought to be treated separately. These are the effects on the conclusions of (a) the spacing of the quadrats and (b) the sizes of the quadrats.

2. The Spacing of the Quadrats

In this section we discuss only positive association. The same arguments, with obvious modifications, apply equally to negative associations.

When two species are found to be positively associated, the conclusion drawn is usually one or both of the following: (a) one of the species has a beneficial effect on the other, either directly or by modifying the environment in a way favorable to it; or (b) some independent environmental factors are variable over the area, and because the two species have identical or overlapping tolerance ranges for the factors, both are forced to occupy coincident or overlapping areas.

The fact that a statistical test gives evidence that two species are positively associated does not, of course, lead directly to the conclusion that one of these mechanisms must be operating. The test by itself suggests only that the null hypothesis should be rejected. The null hypothesis is this: the probability that a quadrat contains species A is independent of whether it does or does not contain species B, and vice versa. Rejection of the hypothesis, and the consequent acceptance of the alternative hypothesis, namely that the probabilities are *not* independent, does not automatically imply that it is the two species that are dependent. It may simply mean that the quadrats are dependent.

Thus an unexpectedly high number of joint occurrences of the two species could easily arise from the following cause. Suppose both species have patchy spatial patterns merely as a result of their modes of reproduction. Then, if, for both species, their patches are large relative to the study area, the extensive overlap of a large patch of species A with another of species B will result in the joint occupancy by both species of a large region of overlap, all this being purely a matter of chance. We need not assume that the species have any beneficial effect on one another, or that the environment is heterogeneous.

Imagine, for concreteness, that each pattern has the form of a random L- or S-mosaic (see Chapter 12). Then, if the two species are independent, their joint pattern will be that of two random mosaics haphazardly superimposed. The joint pattern will be a four-phase mosaic with phases that can be labeled (AB), $(A\bar{B})$, $(\bar{A}B)$, and $(\overline{AB})$. Now, if this mosaic were coarse-grained, that is, if its patches were large in relation to the whole area being sampled, then many of the quadrats placed in it would not be independent of one another. The fact that one quadrat proved to be an (AB), for example, would mean that nearby quadrats also had a high probability of being (AB). Only if the quadrats were so widely separated that their mutual dependence was negligible could we safely conclude that any observed association was not the result of chance overlap.

The association that may be observed when a coarse-grained mosaic is sampled with closely spaced quadrats is not in any sense "unreal," but it is due merely to chance and not to the operation of biological causes. In order to test for the existence of association having a true biological cause it is therefore necessary to space the sample quadrats far enough apart to ensure that there is only a negligible probability that any two of them will occupy the same patch of either species. Unless this is done, chance association is likely to be mistaken for biological association. It should be emphasized that association resulting from patch overlap should be ascribed to chance only if the patches themselves are the outcome of chance. If the patches owe their existence to especially favorable patches of ground, then association resulting from patch overlap has indeed a biological cause.

It is obvious that chance overlap might as easily produce an unexpectedly high proportion of $(A\bar{B})$, $(\bar{A}B)$, or $(\overline{AB})$ quadrats. If the area sampled is large enough and there is no biological association between the species, the localized exesses of quadrats of one kind will tend to cancel one another out. It is these localized excesses that give rise to chance association within small areas. If, in a large area, we still find evidence of positive association, it follows that the two one-species mosaic patterns are not haphazardly superimposed. Then the association must have a biological cause and the overlap is not wholly due to chance.

We may, of course, choose to use closely spaced quadrats in studying a population of small area, but this should be done only when the objective is to study a spatial pattern formed of two superimposed one-species patterns. This is a legitimate endeavor, though it seems not to have been tried to date. Far more often tests of association are prompted by a desire to know if the two species are causally associated. What must then be avoided is the placing of several quadrats in any single mosaic patch.

Having stressed that quadrats must be widely spaced in a test for causal association, the question now arises: how can one tell whether a particular spacing is wide enough? Suppose sampling is done systematically, with quadrats at the corner points of a square lattice (see Figure 22). Now let a map of the quadrat layout be drawn and let the (AB) quadrats be marked and the rest left unmarked. If many of the (AB) quadrats are in a few large regions of patch overlap, they will tend to fall into groups; in other words, they will not be randomly dispersed among the available lattice points. Conversely, if the (AB) quadrats are in different regions of overlap, they will not form groups; instead they will be randomly and independently mingled with the other quadrats. Thus, if we take as a null hypothesis that the (AB) quadrats are independent of one another, we may easily test this hypothesis by means of Krishna Iyer's test (see page 107), applied now to a lattice of spaced quadrats instead of to a grid of contiguous quadrats. The

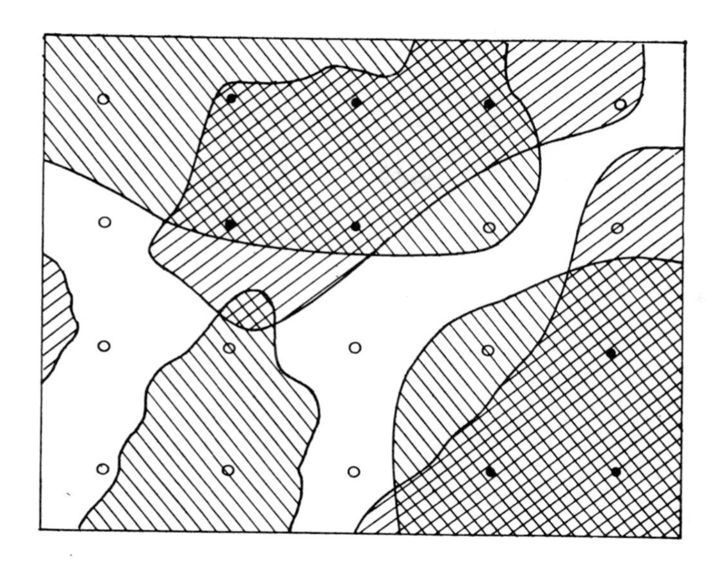

Figure 22*a*. Systematic sampling of a coarse-grained mosaic, with dependence among the quadrats; (AB) quadrats (shown ●) are in groups relative to the other quadrats (shown ○).

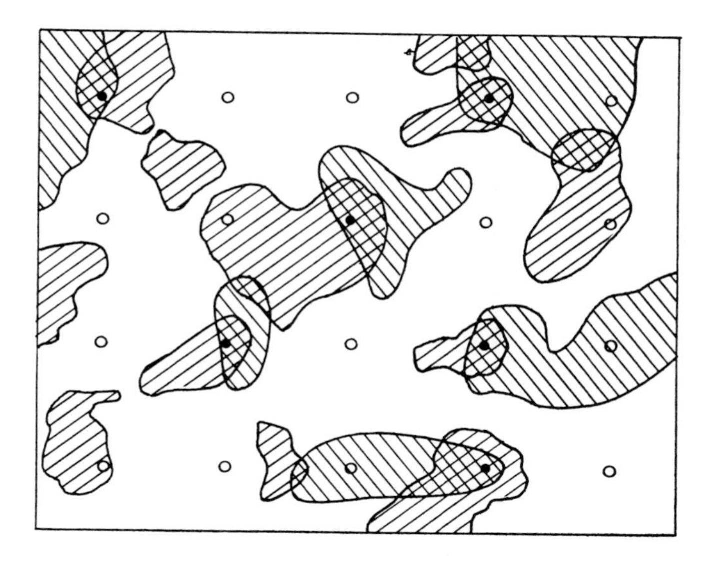

Figure 22*b*. Systematic sampling of a fine-grained mosaic; the quadrats are independent and (AB) quadrats are randomly mingled with the others.

power of the test is probably not great, but its use should at least permit the detection of pronounced dependence among the (AB) quadrats. Judgment by intuition is likely to be unreliable in this context.

If we wish to examine a large sample of widely spaced quadrats, it is clear that the study area itself must be large. It is well known that judgments of association are strongly affected by the size of the area within which sampling is done. Thus suppose we define the population under study as a small area, throughout which both the species are common, and we find no evidence of association between them. If the population is now redefined to include large surrounding areas of ground from which both species are absent and the enlarged area is sampled, the two species will show strong positive association. This must obviously happen, since the proportion of $(\overline{AB})$ quadrats in the large area will greatly exceed the proportion in the small area (see page 170). It seems often to be believed that to include in a sample a great many quadrats that for unknown reasons may be incapable of containing either of the species invalidates an association test in some way. The notion seems to be that the evidence for positive association has been obtained by cheating. This is not so. If large areas of ground contain neither of the species, it is reasonable to conclude that the two species resemble each other in the conditions they will *not* tolerate. This in itself shows that they are associated in the sense that there is considerable overlap in their tolerance ranges. The search for association of this type should obviously not be restricted to small homogeneous areas. If we wish to know whether two species will react similarly to environmental conditions, it is absurd to insist that sampling be confined to a homogeneous area throughout which the conditions are constant. To do so would be to rule out the possibility of finding different responses to different conditions.

Coefficients of association, such as V, Q, and C, described in Chapter 13, are inevitably affected by the amount of ground unsuitable for both the species that happen to be included in a study area. We can only say that a measure of association must always be regarded as a property of a species pair and an area considered *jointly*. In an attempt to overcome this defect in the ordinary association coefficients some workers (e.g., Dice, 1945, and Bray, 1956) have proposed what amounts to a coefficient of overlap, though the different authors use various names and definitions for their coefficients. Denoting the observed frequencies of (AB), $(A\bar{B})$, $(\bar{A}B)$, and $(\overline{AB})$ quadrats by a, b, c, and d, as in a 2×2 table (see page 160), all of these coefficients are functions of a, b, and c only; d is disregarded. Thus, for instance, Bray's (1956) "coefficient of amplitudinal correspondence" is $2a/(2a + b + c)$; that is, it is the ratio of the number of quadrat occurrences of A's and B's, when they occur jointly, to the total number of occurrences of both species.

The use of a coefficient of overlap is, however, no solution to the difficulty just mentioned—that a coefficient of association is strongly influenced by d, the number of empty quadrats. A coefficient of overlap contains no new information: one cannot judge whether it departs significantly from expectation, on the null hypothesis of independence of the species, without taking d into account.

3. The Effect of Quadrat Size

We turn now to a consideration of the effects of quadrat size on indications of association, assuming quadrat spacing to be wide enough for no problems to arise in that connection.

Clearly, only a limited range of sizes is permissible. The quadrats must not be so small that they are incapable of containing at least two individuals of the larger species. Nor must they be so large that one of the two species will occur in every quadrat; this would cause one of the marginal totals of the 2×2 table to be zero and make a test impossible. For practical reasons the feasible range of quadrat size will often lie well within the theoretically permissible range.

Suppose that the two species whose relationship is being studied are associated; that is, they are not independent. Suppose also that this association is caused by the fact that each is responding to the same controlling factor in the environment. For concreteness imagine this factor to be soil moisture. We could, at least in theory, mark the tolerance range of each species on a linear scale of moisture values. Various relationships are possible. The range of one species could be wholly within that of the other; the ranges of the two species might be coincident, they might overlap to a greater or lesser degree, or they might be disjunct but contiguous. Lastly, they might be disjunct and separated by a gap representing soil too dry for one species and too wet for the other.

When two species are controlled in this way by the same factor, they will certainly show association at some quadrat sizes, but the magnitude and even the sign of the observed association must depend on quadrat size. If the tolerance ranges are coincident or overlap markedly, even small quadrats will show positive association, but if the tolerance ranges are disjunct or only slightly overlapped small quadrats may give evidence of negative association and larger quadrats evidence of positive association. The way in which association varies with quadrat size therefore depends on the interrelation between the species' tolerance ranges for the factor and also on the gradient of this factor on the ground; that is, on whether its spatial variation is abrupt or gradual. Thus, if two species showed negative association at a given quadrat size in a region in which the factor varied only gradually, the same two species might be found to be positively associated, at the *same* quadrat size, in a region in which the controlling factor varied abruptly.

It is clear that any observed relation between association and quadrat size is capable of a variety of interpretations. This does not mean that association tests are useless; for example, suppose we were interested in three species, A, B, and C, and found that with a given size of quadrat A and B and also B and C showed marked positive association, whereas for A and C the association was only weakly positive or perhaps negative. We could then conclude that B's tolerance range for some factor lay between those of A and C.

In any case, whenever the association between two species is tested or measured, the size of quadrat used for sampling must always be stated. This is merely a matter of defining the sample units, a thing one would always do whether the units were discrete natural objects or arbitrary pieces of a continuum.

It must again be emphasized that if the tolerance ranges of two species overlap there will be regions on the ground in which the actual overlap of patches of plants may result only from chance. This is because, within a region suitable for it, the pattern of a species is likely to be patchy because of its mode of reproduction. What is observed is not patches of suitable gound but familial patches *within* the patches of suitable ground. (Because of this even total coincidence of two species' environmental requirements may give coefficients of association significantly less than one.) Therefore only if the quadrats are spaced widely enough to encompass unsuitable as well as suitable ground and an extensive area of each can observed association be ascribed to biological causes.

To study association by means of grids of contiguous quadrats, as suggested by Greig-Smith (1964) and Kershaw (1960), can only lead to confounding of two effects, those of quadrat spacing and quadrat size. For investigating association, as well as for investigating pattern, these authors use grids of contiguous quadrats and then combine the quadrats in pairs to form successively larger blocks (see page 104). Thus interquadrat distance and $\sqrt{(\text{quadrat area})}$ are directly proportional and the results become impossible to interpret.

15

Segregation Between Two Species

1. Introduction

It was shown in Chapter 14 that when we examine the association between two species of plants the results will be strongly influenced by both the spacing of the quadrats and their sizes. This is because what is being investigated is not so much interspecies relationships per se but rather joint, two-species patterns. Other factors, besides the relationship between the species, affect these joint patterns. This suggests that it would be worthwhile to attempt to study the pattern of each species in *relation to the other* without regard to the pattern of either in relation to the ground.

We assume that the plants occur as discrete, genetically distinct individuals, reproducing by seed, and that therefore we shall not be misled by the presence of clumps of vegetative shoots which are, or may have been in the past, organically connected. What we now enquire is: do the two species form "relative clumps"? A relative clump of species A, for instance, occurs in a group of plants in which the proportion of A's is greater than their proportion in the whole population. Likewise for species B. A relative clump may or may not be a spatial clump also. Thus consider Figure 23. In Figure 23*a*, although the plants as a whole have a random spatial pattern, there is clear evidence of relative clumping. Conversely, in Figure 23*b*, although the population as a whole is strongly clumped, the two species are not clumped in relation to one another, since within each clump the A's and B's are present in the same proportions and are randomly mingled.

The objective now is to study the relative patterns of two species independently of their spatial patterns. The first question that arises is: are the two species randomly mingled or are they relatively clumped? If they are randomly mingled, they may be described as unsegregated; if not, they are to some extent segregated from each other.

One way of answering this question would be to sample the population with quadrats and adjust the size of each quadrat so that it contained exactly

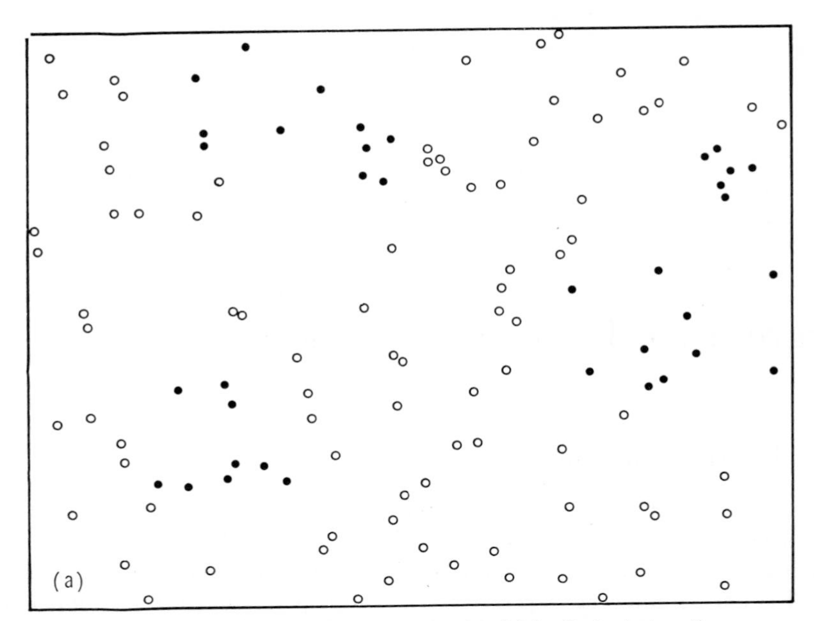

Figure 23*a*. A two-species population in which all plants together have a random pattern, although the species are segregated.

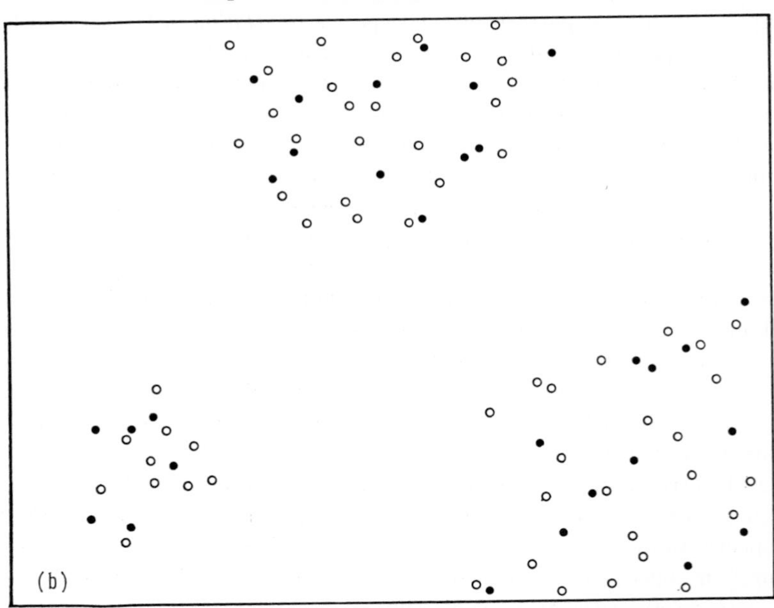

Figure 23*b*. A two-species population in which the plants have clumped patterns but the species are unsegregated.

n individuals. Then, if the two species were randomly mingled and were present in proportions p_A and p_B ($p_A + p_B = 1$), the expected proportion of quadrats containing r A's and $n - r$ B's would be given by the binomial probability

$$\binom{n}{r} p_A^r p_B^{n-r}.$$

If this binomial distribution fitted the observations, we could conclude that the species were randomly mingled. The sampling could be done by taking randomly located points as the centers of circular quadrats and enlarging each circle until exactly n plants were contained in it. As explained on page 119, this is very difficult to do in practice with n greater than about 2 or 3. If n is low, relatively isolated plants will be overrepresented in the sample compared with crowded ones (see page 120).

Alternatively, we might use quadrats of constant size, large enough for the great majority of them to contain at least two plants. If the species were randomly mingled, the results would then consitute a mixture of binomial distributions with several different values of n, the number of plants per quadrat (e.g., see Pielou, 1963a).

Neither of these methods is very satisfactory, and, indeed, quadrat sampling is an inappropriate method for studying relative patterns. A quadrat is, after all, a sample piece of space; it is an appropriate sampling unit if spatial patterns are being studied but not necessarily the best in other contexts. For studying relative patterns it seems more reasonable to use nearest-neighbor methods.

2. Testing for Segregation in a Fully Known Population

The method we shall examine in detail is one that is useful when the population is small enough for all the individuals of both species to be taken into account. Suppose we examine each individual in turn and note its species and that of its nearest neighbor. In judging which neighbor is nearest, distances are measured from center to center of the plants concerned. The results may be set out in a 2×2 table as follows:

		Species of nearest neighbor		
		A	B	
Species of base plant	A	a	b	m
	B	c	d	n
		r	s	N .

A χ^2-test may be used to judge whether the observed cell frequencies depart significantly from expectation. Thus, if $a > mr/N$ significantly, A's have other A's as nearest neighbors unexpectedly often, and the species are partly segregated.

Notice that m and n are the population frequencies of species A and B; they are not estimates, since every plant in the whole two-species population has been examined. The column totals r and s are the numbers of times individuals of species A and of species B have served as the nearest neighbor of another individual. There is no reason to expect r and s to be identical with m and n. It is clear that in any plant population some individuals will be so placed that they are not the nearest neighbor of any other individual; others will serve as nearest neighbors to 1, 2, ..., 5 individuals. (Since distances are measured from center to center, it is impossible for any individual to serve as nearest neighbor to more than five others.) Thus the "population of nearest neighbors" is not the same as the "population of base plants" (i.e., the population in the ordinary sense). The population of nearest neighbors does not contain those base plants that are not the nearest neighbor of any any other plant, whereas the remaining base plants contribute 1, 2, ..., 5 members to the nearest neighbor population according as they are the neighbors of 1, 2, ..., 5 base plants. For this reason we do not expect to find $r = m$ and $s = n$ exactly. If both species have random spatial patterns, the expected proportions of A's and B's in the nearest-neighbor population will be the same as their proportions in the base population; in other words, the column totals will be homogeneous with the row totals. In general, however, even this will not be true.

To see this, suppose the individuals of species A tend to be larger or to require more space than those of species B; that is, the A's tend to be more isolated. The A's will then serve as nearest neighbors relatively less often than the B's and the column totals will not be homogeneous with the row totals. Such a population may be called unsymmetrical.

Random mingling of the species may occur in an unsymmetrical population as well as in a symmetrical one. In testing for random mingling, each plant is treated as having two attributes: its own species and the species of its nearest neighbor. The two species are unsegregated if the two attributes are independent.

There are many ways in which we might define a coefficient of segregation. The one originally suggested (Pielou, 1961) is S, defined as

$$S = 1 - \frac{\text{observed number of mixed pairs}}{\text{expected number of mixed pairs}},$$

where a "mixed pair" denotes an individual of one species having an individual of the other as nearest neighbor.

Then

$$S = 1 - \frac{N(b + c)}{ms + nr}.$$

In an unsegregated population $E(S) = 0$. In a fully segregated in population which no A's have B's as nearest neighbors and vice versa, or $b = c = 0$, $S = +1$.

Negative segregation is also theoretically possible, though in plant populations it is very unlikely. Imagine a population of G widely spaced groups of plants, each consisting of a single central individual of species A with a "halo" of g B's around it. We suppose that each B has the A at the center of its group as nearest neighbor, so necessarily $g \leq 5$. The 2×2 segregation table is

	Species of neighbor: A	B	
Species of base plant A	0	G	G
Species of base plant B	Gg	0	Gg
	Gg	G	N

Then

$$S = 1 - \frac{(G + Gg)N}{G^2 + G^2g^2}$$

$$= 1 - \frac{(1 + g)^2}{1 + g^2} < 0,$$

since $g > 0$. Differentiating with respect to g, we find that S takes its minimum value when $g = 1$, that is, when the whole population consists of isolated A-B pairs; then $S = -1$. Thus S ranges from -1 for the maximum possible negative segregation, through 0 for no segregation, to $+1$ for the maximum possible positive segregation.

If S is determined in the way described above, that is, by inspecting every member of a two-species population, it is a population parameter and has no sampling variance. If the observations are made on a random sample of the population, the calculated S will be only an estimate of the true population value and will be subject to sampling variation. The sampling distribution of S in an unsegregated population has not so far been derived.

The following is an example of an ecological problem on which segregation studies threw light (Pielou, 1966a). Consider a dense stand of seedling trees of two or more species growing up on burned-over land. Initially the trees are likely to be relatively clumped (or partly segregated) for two reasons: (a) the seeds from which they grew may have had clumped patterns and (b) the habitat may be heterogeneous so that one species is favored in one small area and another in another. As the trees grow, their number is bound to decrease due to natural thinning that results from suppression of the less successful trees. If the deaths are preponderantly in crowded one-species clumps that have grown from clumps of seeds, segregation will decrease as the population ages. Conversely, if the trees that die out are those that chanced to germinate in comparatively unsuitable sites, the deaths will tend to sort the species into more clearly defined one-species groups; then segregation will increase. In the work mentioned above the degree of segregation was determined (by a different method from that described here) in five dense stands of young trees at the beginning and end of a period of several years. In all cases segregation was found to decrease, which suggests that natural thinning was the result of intraspecific competition within dense one-species clumps.

3. Plant Sequences in Transects

To identify only the nearest neighbor of a plant in a two-species population will obviously give less information about the relative patterns of the species than would the identification of the 1st, 2nd, ..., nth nearest neighbors. As already explained (page 119), the larger the value of n, the more difficult it is to make the field observations. A way around the difficulty consists in studying the sequence of plants (of the two species concerned) along a belt transect; the transect should be just narrow enough for there never to be any doubt of the order of the plants within it. The resulting observations then consist of a sequence of occurrences such as

$$AAA\ BB\ AAA\ BBBBB\ A\ BBB \ldots.$$

Assuming that we have such an observed sequence, we now wish to test whether the A's and B's are randomly mingled. Suppose there are a A's and b B's. For a short sequence ($a, b \leq 20$) an exact test may be done as follows (Feller, 1968):

Let every uninterrupted sequence of A's, or of B's, be called a *run*. If the observed number of runs is small, it is reasonable to suspect that the species are segregated. Therefore, if the observed sequence contains R runs of both species, we wish to determine the probability of there being R runs or fewer on the null hypothesis of random mingling of the species. If this probability is α, we may assert that the species are significantly segregated at the $100\alpha\%$ level.

We now need to determine the probability $P(k)$ that there will be exactly k runs. Note first that the total number of visibly different arrangements of a A's and b B's is

$$\frac{(a+b)!}{a!b!} = \binom{a+b}{a}.$$

Each of these distinguishable arrangements represents the same number, namely $(a!b!)$, of different *in*distinguishable permutations of the A's and B's. Therefore, since all permutations have equal probability, so also have all the distinguishable arrangements.

Next we must find the number of arrangements of the A's and B's that will give k runs.

Suppose, first, that k is even and put $k = 2m$. Then there are m runs of A's and m runs of B's. Now the number of ways of placing a objects into m cells so that every cell contains at least one object is $\binom{a-1}{m-1}$. (For a proof of this see Feller, 1968.) This is the number of ways in which the A's can be partitioned to give m runs. Likewise, the B's can be partitioned into m runs in $\binom{b-1}{m-1}$ ways. Therefore the number of arrangements that will give rise to m runs of each species when the species of the first run is specified is

$$\binom{a-1}{m-1}\binom{b-1}{m-1}.$$

Then, since a run of either species is equally likely to start the sequence, the number of ways of obtaining m runs of both A's and B's, regardless of which species starts the sequence, is

$$2\binom{a-1}{m-1}\binom{b-1}{m-1}.$$

To determine $P(k)$ we note that

$$P(k) = \frac{\text{number of arrangements of } a \ A\text{'s and } b \ B\text{'s that give } k \text{ runs}}{\text{number of distinguishable arrangements of } a \ A\text{'s and } b \ B\text{'s}},$$

so that, when $k = 2m$,

$$P(2m) = \frac{2\binom{a-1}{m-1}\binom{b-1}{m-1}}{\binom{a+b}{a}}.$$

Next, suppose k is odd and put $k = 2m + 1$. Then either we must have $m + 1$ runs of A's and m runs of B's, which can happen in $\binom{a-1}{m}\binom{b-1}{m-1}$ ways or there must be m runs of A's and $m+1$ runs of B's which can happen in $\binom{a-1}{m-1}\binom{b-1}{m}$ ways. Then

$$P(2m+1) = \frac{\binom{a-1}{m}\binom{b-1}{m-1} + \binom{a-1}{m-1}\binom{b-1}{m}}{\binom{a+b}{a}}.$$

Finally, the probability we wish to find, namely that of obtaining R runs or fewer, is given by $\sum_{k=2}^{R} P(k)$. The summation begins at $k=2$ since obviously there must be at least two runs. These probabilities have been tabulated by Swed and Eisenhart (1943) for all $a, b \leq 20$. In applying the test, therefore, we may simply consult the tables and there is no need to calculate and sum all the separate probabilities.

Outside the range of the tables use of this exact combinatorial test is laborious. For long sequences a different test is desirable. It is easy to devise one from the following considerations.

Each encounter with a plant, as we travel along the transect, may be thought of as a trial, the outcome of which must be A or B. If the species are unsegregated, the trials are independent, and the probabilities of the two possible outcomes are constant along the whole transect. Suppose these probabilities are p_A and $p_B = 1 - p_A$. Then the probability of encountering a run of r A's is $p_A^{r-1} p_B$; similarly, the probability of encountering a run of s B's is $p_B^{s-1} p_A$; that is to say, for each species the run lengths (numbers of individuals in the runs) are geometrically distributed. For the A's the expected run length is

$$E(l_A) = \sum_{j=1}^{\infty} j p_A^{j-1} p_B = \frac{p_B}{(1-p_A)^2} = \frac{1}{p_B},$$

and for the B's the expected run length is

$$E(l_B) = \frac{1}{p_A}.$$

Writing $\bar{l}_A$ and $\bar{l}_B$ for the observed mean run lengths of the species, we may therefore derive estimates $\hat{p}_A$ and $\hat{p}_B$ of the probabilities by putting $\hat{p}_A = 1/\bar{l}_B$ and $\hat{p}_B = 1/\bar{l}_A$. (These are, in fact, maximum likelihood estimators of the population probabilities.) As a test statistic we may therefore use $1/L = 1/\bar{l}_A + 1/\bar{l}_B$ and it is clear that, given the null hypothesis, $E(1/L) = 1$. We shall not explore the sampling variance of $1/L$ here. Details will be found in Pielou (1962b).

4. Distributions of Run Lengths for Three Model Populations

Given a sequence of species occurrences within a transect, we can do more than merely test the species for segregation. It is also interesting to devise models that might "explain" the patterns of the two species in relation to each other and to carry out tests to determine whether the models accord with the observations. A model that does not fit the observations is as instructive in some ways as one that does. The discrepancy between observation and expectation itself constitutes an observation, sometimes a more revealing one than that provided by the original raw data.

The Markov Chain Model

The simplest postulate that can be made is that the sequence of species-occurrences forms a realization of a simple two-state Markov chain. The Markov matrix may be written

$$\begin{pmatrix} p_{AA} & p_{AB} \\ p_{BA} & p_{BB} \end{pmatrix},$$

where p_{AB}, for example, denotes the probability that an A will be followed in the sequence by a B and the other three transition probabilities are defined analogously. Then the probability that a run of A's will be of length r is $p_{AA}^{r-1}p_{AB}$ and the mean run length of the A's is $E(l_A) = 1/p_{AB}$. Similarly, the mean run length of the B's is $E(l_B) = 1/p_{BA}$.

It is seen that the run lengths again have geometric distributions, as was true when the species were assumed to be randomly mingled, but if the sequence is a Markov chain we no longer have that $E(1/L) = 1$. If the species are positively segregated, $p_{AA} > p_{BA}$ and $p_{BB} > p_{AB}$. Therefore, since $p_{AA} + p_{AB} + p_{BA} + p_{BB} = 2$, $p_{AB} + p_{BA} < 1$ and consequently $E(1/L) = E(1/\bar{l}_A + 1/\bar{l}_B) < 1$.

In the few tests that have been done (Pielou, 1962b) on the relative patterns of pairs of plant species it was found that in all cases the observed distributions of run lengths had greater variances than had the fitted geometric distributions. On the scanty evidence that so far exists it seems unlikely that the Markov chain hypothesis will often be tenable. We must therefore search for a distribution whose variance is greater than that of a geometric distribution with the same mean.

The Geometric-Poisson Distribution

We now suppose that the population consists of nonoverlapping clumps of the two species and that the clumps are randomly mingled (see Figure 24).

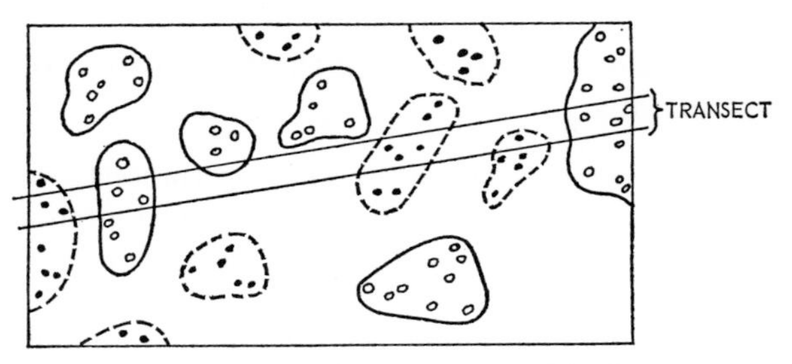

Figure 24. A model population in which the run-lengths of the individuals of each species have Geometric-Poisson distributions. The clumps, shown with solid outlines for one of the species and dashed outlines for the other, are randomly mingled. For each species the number of individuals per clump within the transect is a Poisson variate.

Within a belt transect, therefore, there are alternating runs of A-clumps and B-clumps; (notice that we are now speaking of runs of *clumps*). We do not postulate that there is any visible boundary between adjacent clumps. Since the clumps are assumed to be randomly mingled, their run-lengths must be geometrically distributed.

We next suppose that for each species the number of individuals per clump (within the transect) is a Poisson variate.

For clarity of exposition we now confine attention to the run-lengths of individuals of only one of the species, say species A. Our initial aim is to find the relation between the mean and variance of the distribution, since what is required is a distribution with variance greater than that of a geometric distribution of the same mean. If a distribution has a known probability generating function (pgf), its mean and variance are easily determined directly; there is no need to derive an explicit expression for the probability of encountering a run of r A's.

Now, since the run-lengths of the *clumps* of A's are geometrically distributed, their pgf is

$$G(z) = qz + pqz^2 + p^2qz^3 + \cdots + p^rqz^{r+1} + \cdots$$

$$= \frac{qz}{1 - pz}.$$

For later reference we note here that the mean M and variance V of the geometric distribution are $M = 1/q$ and $V = p/q^2$, so that

$$V = M^2 - M. \tag{15.1}$$

The number of individual A's per A-clump is a Poisson variate and its pgf is therefore (see page 82)

$$g(z) = e^{\lambda(z-1)}, \quad \text{with} \quad \lambda > 0.$$

Thus the pgf of the run-lengths of *individual* A's is

$$H(z) = G[g(z)] = \frac{qe^{\lambda(z-1)}}{1 - pe^{\lambda(z-1)}}.$$

The distribution with this pgf may be called Geometric-Poisson. It is a generalized distribution (see page 83). Its mean is

$$\mu_1' = H'(1) = \frac{\lambda}{q} \tag{15.2}$$

and the second moment *about the origin* is

$$\mu_2' = H''(1) + H'(1) = \frac{\lambda^2}{q^2}(2 - q) + \frac{\lambda}{q}. \tag{15.3}$$

Note, however, that some A-clumps (those with 0 individuals) will be unobservable. Some runs of A-clumps may consist entirely of these empty clumps and the whole run will then be unobservable. Thus the observed distribution will be truncated with the zero class missing. If the proportion of these empty runs is π_0, the first and second moments about the origin of the truncated distribution will be m_1' and m_2', say, where

$$m_1' = \frac{\mu_1'}{1 - \pi_0} \quad \text{and} \quad m_2' = \frac{\mu_2'}{1 - \pi_0}.$$

Now

$$\pi_0 = H(0) = \frac{qe^{-\lambda}}{1 - pe^{-\lambda}}$$

and

$$\frac{1}{1 - \pi_0} = 1 + \frac{qe^{-\lambda}}{1 - e^{-\lambda}}.$$

Therefore, using (15.2) and (15.3), the first two moments of the observed (truncated) distribution are

$$m_1' = \frac{\lambda}{q}\left(1 + \frac{qe^{-\lambda}}{1 - e^{-\lambda}}\right) \tag{15.4}$$

and

$$m_2' = \frac{\lambda}{q}(2 - q)m_1' + m_1'. \tag{15.5}$$

We next eliminate q from these equations and find a relation between m_1' and m_2' on the one hand, and λ on the other. From (15.4) we obtain

$$2m_1' = \frac{2\lambda}{q} + \frac{2\lambda e^{-\lambda}}{1 - e^{-\lambda}}$$

and from (15.5)

$$\frac{m_2'}{m_1'} = \frac{2\lambda}{q} - \lambda + 1,$$

whence

$$2m_1' - \frac{m_2'}{m_1'} + 1 = \frac{2\lambda e^{-\lambda}}{1 - e^{-\lambda}} + \lambda = \frac{\lambda(1 + e^{-\lambda})}{1 - e^{-\lambda}}. \tag{15.6}$$

Recall from (15.1) that we require a distribution for which the first two moments are related by the inequality

$$m_2 = m_2' - m_1'^2 > m_1'^2 - m_1',$$

or, equivalently,

$$2m_1' - \frac{m_2'}{m_1'} + 1 < 2. \tag{15.7}$$

We now ask whether this requirement is compatible with (15.6). The answer is obtained by considering the properties of the function

$$f(\lambda) = \frac{\lambda(1 + e^{-\lambda})}{1 - e^{-\lambda}}.$$

Using l'Hôpital's rule to evaluate $f(0)$ gives

$$f(0) = \lim_{\lambda \to 0} \frac{\lambda(1 + e^{-\lambda})}{1 - e^{-\lambda}} = \lim_{\lambda \to 0} \frac{1 + e^{-\lambda} - \lambda e^{-\lambda}}{e^{-\lambda}} = 2.$$

Further, by differentiating $f(\lambda)$ it is easy to show that $f'(\lambda) > 0$ for $\lambda > 0$; in other words, for $\lambda > 0$ the function is monotonically increasing.

It follows that the Geometric-Poisson distribution does *not* meet the requirement stated in (15.7). On the contrary, the Geometric-Poisson must always have a variance *less* than that of a simple geometric distribution of the same mean. It will therefore fit the observed frequency distributions less well than the simple geometric series.

In postulating that the run-lengths of the A-clumps had a geometric distribution, we took it for granted that the form of the distribution would be unaltered by the fact that some B-clumps, hence some runs of B individuals, would contain 0 individuals. The runs of A's preceding and following such an empty run of B's will, of course, appear to be a single unbroken run. This does not affect the form of the distribution of the run-lengths of A-clumps; it merely changes the parameter of the geometric distribution. Thus, if all B-clumps were observable and the probability of encountering one were p_B, we should have, as the probability of observing a run of r A-clumps, $p_A^{r-1}p_B$ (where $p_A = 1 - p_B$) (cf. page 186). Now let the number of individuals per B-clump be a Poisson variate with parameter λ_B. Then the probability of encountering an *observable* B-clump will be $p_B(1 - e^{-\lambda_B}) = p'_B$, say. Putting $p'_A = 1 - p'_B$, the probability of observing a run of r A-clumps now becomes $(p'_A)^{r-1}p'_B$ and is seen to be still a geometric term.

The Compound Geometric Distribution

Consider again the run lengths of one of a pair of segregated species. As a third possible model, we shall suppose that the individuals of this species have run lengths that are geometrically distributed but that the parameter p of the geometric distribution is itself a random variable. It is assumed that p has a constant value within any one run but different values in different runs. The resulting distribution therefore is compound (see page 86).

We now choose a standard distribution that is likely to provide an approximation to the true distribution of p, whatever form the latter may take. The Beta distribution with range 0 to 1 is suitable because of its flexibility; it may be bellshaped or U-shaped and of any degree of skewness. We therefore assume that the pdf of p is

$$f(p) = \frac{1}{B(\alpha, \beta)} p^{\alpha-1}(1 - p)^{\beta-1},$$

with $\alpha > 0$ and $\beta > 2$. The constraint $\beta > 2$ is necessary to ensure the convergence of the integral in (15.9).

We now wish to find the mean and variance of this compound geometric distribution to see how they are related.

For the simple geometric distribution the mean is $1/q$. Therefore the mean of the compound geometric distribution is

$$\mu'_1 = \frac{1}{B(\alpha, \beta)} \int_0^1 \frac{1}{q} \cdot p^{\alpha-1}(1 - p)^{\beta-1}\, dp = \frac{\alpha + \beta - 1}{\beta - 1}. \tag{15.8}$$

Next, for the simple geometric distribution the second moment about the origin is $(1 + p)/q^2$. Hence the corresponding moment for the compound geometric is

$$\mu'_2 = \frac{1}{B(\alpha, \beta)} \int_0^1 \frac{1 + p}{q^2} p^{\alpha-1}(1 - p)^{\beta-1}\, dp = \frac{(\alpha + \beta - 1)(2\alpha + \beta - 2)}{(\beta - 1)(\beta - 2)}. \tag{15.9}$$

(Notice that, as already remarked, we must have $\beta > 2$ if the variance of the distribution is to be finite.)

Then the variance of the compound geometric is

$$\mu_2 = \mu_2' - \mu_1'^2 = \frac{\alpha\beta(\alpha+\beta-1)}{(\beta-1)^2(\beta-2)}. \tag{15.10}$$

Again recall that we are searching for a distribution with variance greater than that of a simple geometric distribution with the same mean. The compound geometric meets this requirement. From (15.8) and (15.10) it is seen that

$$\mu_2 = \frac{\alpha+\beta-1}{\beta-1} \cdot \frac{\alpha\beta}{(\beta-1)(\beta-2)} > \frac{\alpha+\beta-1}{\beta-1} \cdot \frac{\alpha}{\beta-1}, \qquad \text{since } \beta > 2.$$

Therefore

$$\mu_2 > \frac{\alpha+\beta-1}{\beta-1}\left(\frac{\alpha+\beta-1}{\beta-1} - 1\right) = \mu_1'^2 - \mu_1'. \tag{15.11}$$

This is in contrast to the relation given in (15.1) for the moments of the simple geometric, namely, $\mu_2 = \mu_1'^2 - \mu_1'$.

The probability that a run of individuals will be of length r is given by

$$\pi_r = \int_0^1 p^{r-1}(1-p)f(p)\,dp = \frac{1}{B(\alpha,\beta)}\int_0^1 p^{\alpha+r-2}(1-p)^{\beta}\,dp$$
$$= \frac{B(\alpha+r-1,\beta+1)}{B(\alpha,\beta)},$$

whence

$$\pi_1 = \frac{\beta}{\alpha+\beta}; \quad \pi_2 = \frac{\alpha}{\alpha+\beta+1}\pi_1; \ldots; \pi_r = \frac{\alpha+r-2}{\alpha+\beta+r-1}\pi_{r-1}; \ldots.$$

As shown by (15.11), it is always possible to find a compound geometric distribution with the same mean and variance as an observed distribution of run-lengths, however great the variance of the latter may be. This, of course, does not ensure a good fit of the theoretical distribution to the observed distribution. When the compound geometric was fitted to run lengths of plants observed in the field (Pielou, 1962b), it was found that the number of runs of unit length (i.e., consisting of a single individual) always exceeded expectation. This result suggests that even when two species of plant are unquestionably clumped relative to each other there is often considerable clump overlap. When this is so, it seems reasonable to conclude that the tolerance ranges of the species for microclimatic and edaphic factors must also overlap.

16

Segregation Among Many Species: n-Phase Mosaics

1. Introduction

The methods described in Chapter 15 for studying what may be called segregation patterns in two-species populations can obviously be extended to many-species populations. These methods presupposed that the plants under study occurred as separate, easily distinguished individuals; they are therefore most likely to be applicable in investigations of the relative patterns of the different tree species in mixed forest. It is unrewarding to contemplate many-species segregation patterns in the abstract when one has no particular concrete problem in mind requiring solution, and no attempt is made here to generalize the preceding arguments. However, the concepts of relative pattern and spatial segregation should be borne in mind whenever the pattern of a many-species tree population poses some definite ecological problem (for an example see Pielou 1963b, 1965b).

Entirely different methods of studying many-species patterns are needed when the plants occur, not as distinct, discrete individuals, but as clumps or patches of appreciable area. The vegetation of swamps or of heaths or moors are examples. In drawing a map of such vegetation it would be impossible to represent individual plants by dimensionless dots, each marking a plant's center, as could be done with a map of a forest; instead the vegetation must necessarily be mapped as a many-phase mosaic, one phase of which might be bare ground.

There is an important contrast between dot maps and mosaic maps that should be emphasized at the outset. A dot map is made up of two kinds of entity: the dots, representing individual plants of various species and the undifferentiated background that constitutes the only continuous phase. (In practice the " background " may be bare ground or any form of low vegetation the ecologist is unconcerned with.) In a mosaic map every point in the mapped area is assigned to one phase or another and all the phases are present as patches of finite extent. Bare ground, if present, is most easily treated as an extra phase; the number of phases is then one more than the number of species. It will be remembered that in considering two-phase mosaics (see Chapter 12) we were considering the pattern of a single plant species; the phases consisted of ground occupied, and not occupied, by the species concerned.

In this chapter we consider many-species mosaics from a theoretical point of view. The practical difficulties that may be encountered in field work can be dealt with only as they arise. It is assumed that there are no discrete "dimensionless" plants in the vegetation (or if there are they are to be ignored) and that every point in the area can be assigned to one and only one phase so that the patches of the various phases do not overlap. We are assuming, in short, that the many difficulties inherent in mapping vegetation have already been overcome and that the object of study is a finished mosaic map.

The arguments to be developed apply to all mosaic maps, not merely to those of small areas of ground in which each separate species of plant is distinguished and mapped. Other kinds of mosaic map also interest ecologists and a few examples were given on page 141. To that list may be added maps showing the outlines of recognizably distinct plant communities, or physiognomically defined vegetation types, or areas of ground differing among themselves in physical and chemical characteristics (e.g., see Pielou, 1965a).

2. n-Phase L-Mosaics

The first question to ask concerning an n-phase mosaic map is: is it random? As remarked on page 141, the question cannot be answered as it stands. The term "random," as applied to mosaics, may be variously defined and the question is unanswerable until the exact type of randomness contemplated has been stipulated. We have therefore to envisage, and specify, a random n-phase mosaic. How could one be constructed?

One obvious way is to draw a network of random cells, either by the random-lines process (page 141) or by the random-sets process (page 145). Now suppose that the mosaic is to have n colors, or phases, in proportions $a_1, a_2, \ldots, a_n$. The cells of the network are independent, and for each of them the probability that it will be of the ith color is $a_i (i = 1, 2, \ldots, n)$. The cells are colored accordingly, and the resulting groups of one or more contiguous cells of the same color constitute the patches of the mosaic.

In what follows we consider only n-phase random lines mosaics (L-mosaics) because of their mathematical tractability. By generalizing the argument in Chapter 12 (page 142) it is easily seen that the sequence of phases encountered along a row of equidistant points across such a mosaic is a realization of a simple n-state Markov chain; but if an observed sequence is found to form a Markov chain it does not follow that the mosaic being sampled can be regarded as random. Another condition must also be met. For a mosaic to be considered random it is obviously necessary that the different kinds of patches should be randomly mingled, and this requires that the Markov matrix should be of the form

$$\mathbf{P} = \begin{array}{c} \\ x_1 \\ x_2 \\ \\ x_n \end{array}\!\!\begin{array}{c} \begin{array}{cccc} x_1 & x_2 & & x_n \end{array} \\ \begin{pmatrix} P_1 & p_2 & \cdots & p_n \\ p_1 & P_2 & \cdots & p_n \\ \cdots & \cdots & \cdots & \cdots \\ p_1 & p_2 & \cdots & P_n \end{pmatrix} \end{array}. \tag{16.1}$$

Here $x_1, x_2, \ldots, x_n$ denote the different phases (or species or colors) occurring along the sequence.

In this matrix all the off-diagonal elements in any one column are equal. In other words, the probability that any of the not-x_1's, say, will be succeeded by an x_1 is the same for all not-x_1's and is given by p_1.

Two properties of a Markov chain having a matrix of this form should be noted.

1. The chain is reversible; that is, if any two adjacent points in the sequence belong to phases x_j and x_k, the two possible orders $x_j x_k$ and $x_k x_j$ will have equal probabilities. Denoting the limiting vector of the chain by $\mathbf{a}' = (a_1\ a_2\ \cdots\ a_n)$, we find that the condition for reversibility is therefore $a_j p_k = a_k p_j$. That this is so may be seen from the following:

Since $\mathbf{a}'\mathbf{P} = \mathbf{a}'$, we may find the elements of $\mathbf{a}'$ in terms of the elements of $\mathbf{P}$ from the equations

$$a_j P_j + P_j \sum_{k \neq j} a_k = a_j, \qquad j = 1, 2, \ldots, n,$$

or

$$p_j(1 - a_j) = a_j(1 - P_j),$$

whence

$$a_j = \frac{p_j}{1 - P_j + p_j}. \tag{16.2}$$

Then the condition $a_j p_k = a_k p_j$ becomes

$$\frac{p_j p_k}{1 - P_j + p_j} = \frac{p_k p_j}{1 - P_k + p_k}$$

or

$$p_j + P_k = p_k + P_j. \tag{16.3}$$

Both members of (16.3) are equal to $1 - \sum_{l \neq j, k} p_l$ and the condition for reversibility is fulfilled.

Vegetation mosaics that might yield irreversible chains are easy to visualize. Suppose the phases of a mosaic consisted of different plant communities and their arrangement was controlled by the prevailing wind so that one type of community was often found downwind but seldom upwind of another type (e.g., see Watt, 1947). The sequence of phases observed along a row of points parallel with the wind direction would then depend on whether the sequence was observed upwind or downwind; that is, the chain would be irreversible. As another example, consider the sequence of phases likely to be observed along a north-south transect in temperate latitudes; the sequence would be irreversible if some small plants tended to grow only on the shaded side and others only on the sunlit side of clumps of taller vegetation. Mosaics such as these are obviously nonrandom.

2. A chain having a matrix of the form **P** in (16.1) is *lumpable* with respect to any partition of the phases (see Kemeny and Snell, 1960); that is, if the phases are grouped together in any manner whatever to give a smaller number of more broadly defined "lumped phases," the sequence will still form a Markov chain. To see this we may argue as follows: suppose the phases $x_1, x_2, \ldots, x_n$ are partitioned into sets $\{X_1, X_2, \ldots\}$. Each set consists of one or more phases lumped together. Now consider the sets X_u and X_v having s and t phases, respectively. Let the phases be so labeled that X_u consists of the phases $x_1, x_2, \ldots, x_s$, and let X_v consist of the phases $x_{s+1}, x_{s+2}, \ldots, x_{s+t}$. Denote by $P(x_i X_v)$ with $i = 1, 2, \ldots, s$ the probability of a transition from the phase x_i in the set X_u to any phase in the set X_v. Then the condition for lumpability of these phases is that

$$P(x_1 X_v) = P(x_2 X_v) = \cdots = P(x_s X_v);$$

that is, the probability of a transition from any phase in the set X_u to a phase in X_v is the same for all phases in X_u. It will be seen that

$$P(x_i X_v) = p_{s+1} + p_{s+2} + \cdots + p_{s+t} \quad \text{for} \quad i = 1, 2, \ldots, s,$$

and therefore the chain is lumpable with respect to any partition of the sets we care to make.

From this it follows that the two-phase mosaic formed of any one phase and all the others combined is a random two-phase L-mosaic of the sort described in Chapter 12. Therefore each phase, or species, considered alone has a mosaic pattern that is random in the sense already defined.

Random n-phase L-mosaics of vegetation are probably rare in nature. Presumably they are as uncommon as many-species populations of "discrete" plants in which every species has a random dot pattern.

It is interesting to compare the number of parameters needed to specify completely (a) the pattern of an n-species population of discrete plants in which all n species are at random and (b) a random n-phase L-mosaic. Clearly (a) is completely defined when the densities, or Poisson parameters, of all the species are known; that is, there are n parameters. Likewise, to specify (b), n parameters are needed as now shown. First it is necessary to know the relative areal proportions of the phases or equivalently the elements of the limiting vector $\mathbf{a}'$; the vector has n elements, but since they sum to unity only $n-1$ are independent. In addition one of the transition probabilities must also be known to complete the specification. This is equivalent to asserting that if $\mathbf{a}'$ is known only one of the elements in $\mathbf{P}$ is free to vary, a fact we shall now prove.

It has already been shown [see (16.2)] that a_j is a function of p_j and P_j only. Thus, when a_j is given, p_j is uniquely determined by P_j or vice versa. It was also shown in (16.3) that $P_j - p_j = P_k - p_k$ or that $P_j - p_j$ is constant for all j. It follows that if all the elements of $\mathbf{a}'$ and one of the elements of $\mathbf{P}$ are given all the other elements of $\mathbf{P}$ will be determined. Thus, of the n parameters needed to define the random mosaic, $n-1$ are determined by the areal proportions of the n phases and the nth is a measure of the "grain" of the mosaic. If $P_j - p_j$ (which is constant for all j) is large, the mosaic is coarse-grained; conversely, if $P_j - p_j$ is small, the mosaic is fine-grained.

3. Unsegregated Mosaics

If, in studying a mosaic, we take as null hypothesis that it has the form of a random L-mosaic we are, of course, testing a very complicated hypothesis. We are assuming not only that the phases are randomly mingled but also that the pattern of each species considered by itself is a two-phase random mosaic. We shall next consider the properties of an *unsegregated mosaic*, or one in which the phases are randomly mingled, although no restrictions are placed on the patterns of the separate species. We are now concerned only with the *relative* arrangements of the different kinds of patches and not with their absolute patterns. In discussing the patterns of discrete "point" plants in Chapter 15, we emphasized the distinction between relative patterns (the patterns of species relative to one another) and absolute patterns (the spatial pattern of each species relative to the ground). We now take an analogous approach to mosaic maps. A random n-phase L-mosaic may be thought of as the analog of the pattern formed by the superposition of n independent random dot patterns. Similarly, an unsegregated mosaic, in which the different kinds of patches are randomly mingled, is the analog of a dot pattern of the kind shown in Figure 23*b* (see page 180), in which the different kinds of dots are randomly mingled, although their spatial patterns need not be random. Obviously the patches in a mosaic can be randomly mingled or unsegregated regardless of their shapes and sizes, and, if a vegetation mosaic is found to be unsegregated, we may conclude that there is no tendency for particular groups of species to grow adjacent to one another.

To investigate the properties of a mosaic that is random in this less restricted sense it is again convenient to consider the sequence of phases encountered at equidistant points along a line transect. Labeling the n different species with the letters A, B, ..., N, we obtain sequences such as the following:

$$AA\ CCCC\ N\ AA\ C\ BBB\ CC\ DDD\ A\ NNN\ \ldots.$$

The sequence is a succession of runs of the different letters. The lengths of these runs is now of no interest and the observations can be replaced by a "collapsed" sequence in which each run, whatever its length, is replaced by a single letter. The collapsed sequence corresponding to the observed sequence shown above is thus

$$A\ C\ N\ A\ C\ B\ C\ D\ A\ N\ \ldots.$$

We shall now show that if the species are randomly mingled this sequence constitutes a Markov chain and shall find its transition probabilities in terms of the elements of its limiting vector.

The argument is clarified by envisaging a balls-and-boxes model that generates a sequence of the desired kind. The following is such a model:

Take n boxes. Put a supply of balls of n different kinds, labeled A, B, ..., N, in each box; these balls are present in the same proportions in every box, namely $\pi_1, \pi_2, \ldots, \pi_n$. Now take the jth box and remove from it all balls bearing the jth letter, for $j = 1, 2, \ldots, n$. The jth box will now contain balls of all letters except the jth; in the jth box the proportion of balls bearing the kth letter becomes $\pi_k/(1 - \pi_j)$ for $k = 1, 2, \ldots, (j-1), (j+1), \ldots, n$. The sequence we require can now be generated as follows: pick a box at random and from it take a randomly selected ball, observe its letter, and replace it. If it had the fth letter, pick the next ball from the fth box; observe the letter of this second ball and replace it; if the second ball bore the gth letter, pick next from the gth box; and so on. In the resulting sequence of letters clearly no two adjacent ones can be the same. It is also clear that at each step of the process the probability of picking a ball with a particular letter depends only on the letter of the preceding ball, since this ball determined which box the succeeding ball should be picked from. Hence the sequence is Markovian. The Markov matrix of the collapsed chain is given by

$$\mathbf{Q} = \{q_{jk}\} \quad \text{where} \quad q_{jk} = \begin{cases} \dfrac{\pi_k}{(1-\pi_j)} & \text{when} \quad j \neq k, \\ 0 & \text{when} \quad j = k. \end{cases}$$

Denote the chain's limiting vector by

$$\mathbf{b}' = (b_1\ b_2\ \cdots\ b_n).$$

Then, since $\mathbf{b}'\mathbf{Q} = \mathbf{b}'$,

$$\sum_{r \neq j} \frac{b_r}{1 - \pi_r} = \frac{b_j}{\pi_j}$$

Therefore

$$\sum_{r \neq j} \frac{b_r}{1 - \pi_r} - \frac{b_j}{\pi_j} = 0 = \sum_{r \neq k} \frac{b_r}{1 - \pi_r} - \frac{b_k}{\pi_k},$$

whence

$$\frac{b_k}{1 - \pi_k} - \frac{b_j}{\pi_j} = \frac{b_j}{1 - \pi_j} - \frac{b_k}{\pi_k}$$

and thus

$$\frac{b_j}{b_k} = \frac{\pi_j(1 - \pi_j)}{\pi_k(1 - \pi_k)}.$$

It follows that $b_j = C\pi_j(1 - \pi_j)$, where C is a constant of proportionality.

The π's may now be found in terms of the b's; computational details will be found in Pielou (1967b). As estimates of the b's, we may take the observed proportions of A's, B's, C's, etc., in the collapsed chain. Notice that these proportions are not the same as the areal proportions of the different phases of the mosaic. They are the proportions of times that *runs* (of any length) of the various phases occur in the original uncollapsed sequence.

If some of the species (or phases) of the mosaic occur only as occasional, widely scattered patches, we may wish to pool two or more of these infrequent species into a single miscellaneous class. It can be shown (Pielou, 1967b) that such pooling will not affect the form of the chain. This is not the same as proving that the chain is lumpable (see page 196) which, in general, it is not. Thus suppose we are given the collapsed chain

$$A\ C\ B\ A\ X\ C\ A\ B\ C\ X\ Y\ X\ C\ B\ C\ A\ Y\ X\ A \ldots$$

and wish to pool the two uncommon species X and Y, calling the combined "species" Z. The sequence obtained by replacing all the X's and Y's by Z's is

$$A\ C\ B\ A\ Z\ C\ A\ B\ C\ Z\ Z\ Z\ C\ B\ C\ A\ Z\ Z\ A \ldots.$$

However, this is *not* the sequence we are interested in, since it no longer has the property that adjacent pairs of letters are always different; it contains runs of consecutive Z's.

Therefore let us collapse the last sequence by deleting all but one Z wherever a run of two or more of them occurs. The result may be called a twice-collapsed sequence, and in this example it is

$$A\ C\ B\ A\ Z\ C\ A\ B\ C\ Z\ C\ B\ C\ A\ Z\ A \ldots.$$

This twice-collapsed sequence is Markovian and its matrix of transition probabilities is of the same form as **Q**. The proof entails showing that the twice-collapsed sequence is identical with the chain that would have been obtained had the pooling been done at the balls-in-boxes stage of constructing the model; that is, if we had used only $(n - 1)$ boxes and had replaced the X-balls and Y-balls with Z-balls.

It follows that if an n-phase mosaic has randomly mingled patches the mosaic formed by combining some phases to form fewer, more broadly defined "pooled phases" also has randomly mingled patches. This fact allows us to combine rare species (or rare phases of any kind) into more inclusive phases in any way that may be convenient before doing a test for random mingling.

IV

Many-Species Populations

17

Species-Abundance Relations

1. Introduction

Most ecological communities contain many species of organisms, and the species may vary greatly in their abundance from very common to very rare. Therefore, as soon as one attempts to study whole communities, rather than the interrelations among a few chosen species, the question immediately arises: how are the abundances of the different species distributed? If there are N individuals belonging to s species and the numbers of individuals in the respective species are $N_1, N_2, \ldots, N_s$, have the N_j any consistent interrelationship, regardless of the type of community from which they come? Attempts to answer this question have led to the development of "species-abundance" curves. If it should turn out that one single form of probability distribution with a small number of parameters (say, two or three) fitted the data from the majority of observed communities, with only the parameter values varying from one community to another, interesting relationships might be discovered between the values of the parameters and the types of community they described.

Before delving into the mathematics, it is necessary to contemplate the word "community" and what is meant by it. In the present context it means all the organisms in a chosen area that belong to the taxonomic group the ecologist is studying. The chosen area is usually one that the ecologist regards as a convenient entity and is willing to consider as homogeneous in some intuitive sense. The reliance on intuition is necessary, since homogeneity cannot be precisely defined at present; exactly what meaning, if any, should be attached to the term "homogeneous community" has for many years been hotly debated and no end to the discussion is in sight. Community studies would have to be suspended indefinitely if ecologists refrained from investigating any community until a satisfactory definition of the word "homogeneous" had been attained. Delimitation of an area that the community under study is supposed to occupy is therefore nearly always a matter of common sense and convenience.

The same is true when it comes to defining the group of animals or plants that is to constitute the community. To take all the living things in the specified area will not do. It would be impracticable to consider every kind of living thing in, say, an acre of forest—the mammals, birds, reptiles, amphibians, arthropods, and soil microfauna, together with the trees, shrubs, herbs, ferns, mosses, and bacteria. A taxonomic group that the ecologist regards as an entity is usually chosen; often it is an entity only in the sense that it is a family, order, or class (or other taxon) that taxonomists are familiar with so that individuals can be fairly easily identified to species.

Thus two subjective choices must be made in defining a community (more precisely, a collection): the taxon, of whatever rank, whose individuals are to be collected and the area or volume within which the collecting is to be done. Examples (from Williams, 1964) are the breeding bird pairs in a tract of forest; the herbaceous vascular plants growing on a homogeneous area of ground, the fresh-water algae in small ponds, the beetles in flood refuse on river banks, the snakes in a stretch of tropical forest, and the moths caught in a light trap. In the last example the exact area involved is, of course, unknown; the collection studied consists of those moths that are phototaxic, within range of the light, and happen to see it.

Without attempting either to justify or to disparage the types of collection examined in attempts to determine their species-abundance relations, we now proceed with the mathematical theory that has been developed to account for them. In many collections it is found that singleton species (those represented by one individual) are numerous, often the most numerous. Species with successively more representatives, 2, 3, . . . , and so on, are usually progressively less numerous. Roughly speaking, one often finds many rare species and a few abundant ones, although, of course, in terms of numbers of individuals those of the few common species far outnumber those of the many rare species. This frequently observed phenomenon has led to the method of tabulating species-abundance data customarily used: instead of listing the numbers of individuals in species 1, species 2, etc., we list the number of *species*, n_1, represented by one member, . . . , the number of species, n_r, represented by r members, . . . , and so on. The n_r are, in fact, frequencies of frequencies. Figure 25 is an example.

In all cases the collection at hand is treated as a random sample from some indefinitely large parent population. Assume further that each species is randomly dispersed; that is, the number of members the collection contains of, say, the jth species is a Poisson variate with parameter λ_j. Then

$$\Pr(\text{the } j\text{th species is represented by } r \text{ members}) = e^{-\lambda_j} \frac{\lambda_j{}^r}{r!}.$$

Now consider all the species in the community. Their densities vary from species to species over a wide range. If there are S^* species in the whole

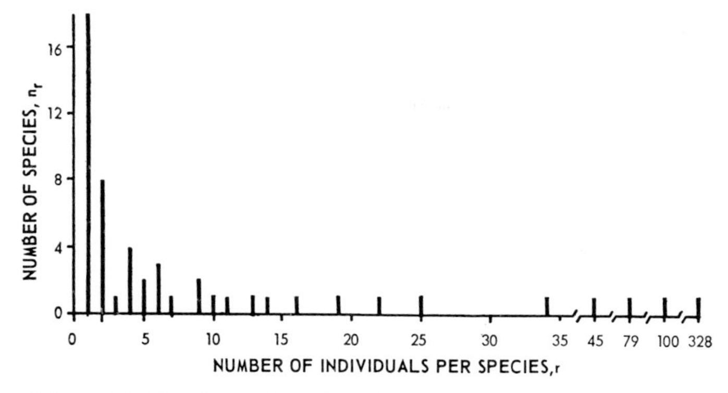

Figure 25. Numbers of species with 1, 2, 3, ..., individuals in a collection of 822 individuals (52 species) of insects and mites. The individuals were adults emerging from a collection of fruiting bodies of the bracket fungus *Fomes fomentarius*. (Data from Pielou and Matthewman 1966).

population, we may regard the several values of λ as constituting a sample of size S^* from some continuous distribution of λ values having pdf $f(\lambda)$. Then the probability that any species will be represented by r members is

$$p_r = \int_0^\infty \frac{\lambda^r e^{-\lambda}}{r!} f(\lambda)\, d\lambda \quad \text{for} \quad r = 0, 1, 2, \ldots; \tag{17.1}$$

that is, the distribution of the different species frequencies n_0, n_1, n_2, ..., where $n_r = S^* p_r$ is assumed to have the form of a compound Poisson distribution (see page 86).

The observed distribution is a truncated form of the theoretical distribution; the zero class is missing. We do not, in general, know the value of S^*, the number of species in the whole population, for presumably some of them will be missing from the collection, which is only a sample of the population. Suppose the observed number of species is S. Then $S^* - S = n_0$ is the number of species represented by zero members in the collection, which is to say unrepresented.

It is worth contrasting this situation with that obtaining when the spatial pattern of one species (e.g., of plant) is being investigated by quadrat sampling. In the latter case we count the numbers of individuals of the one species concerned in a known number of different quadrats located at different places; thus we can count the number of empty quadrats (those from which this species is absent) and so obtain an empirical value of n_0. In obtaining the empirical distribution of species-abundances in a collection, on the other hand, we are examining only a single area (equivalent to one quadrat) and counting the numbers of members it contains of each of S^* different species; since S^* is unknown, so also is n_0.

We now consider those members of the family of compound Poisson distributions that have been fitted to observed species-abundance data.

2. The Logarithmic Series

Suppose the values of λ for the different species are assumed to have a Type III distribution; that is, $f(\lambda)$ in (17.1) is given by

$$f(\lambda) = \frac{P^{-k}\lambda^{k-1}e^{-\lambda/P}}{\Gamma(k)}.$$

Then (see page 87) p_r is a negative binomial variate, or

$$p_r = \frac{\Gamma(k+r)}{r!\,\Gamma(k)}\left(\frac{P}{1+P}\right)^r\left(\frac{1}{1+P}\right)^k \quad \text{for} \quad r = 0, 1, 2, \ldots .$$

It is convenient to write $P/(1+P) = X$ so that

$$p_r = \frac{\Gamma(k+r)}{r!\,\Gamma(k)}(1-X)^k X^r \quad \text{with} \quad 0 < X < 1.$$

Next consider p_r', the probability that a species will contain r individuals, given that the zero class is ignored; that is, p_r' is a term of a truncated negative binomial distribution.

Then, since

$$p_0 = (1-X)^k, \tag{17.2}$$

$$p_r' = \frac{p_r}{1-p_0} = \frac{\Gamma(k+r)}{r!\,\Gamma(k)}\frac{(1-X)^k X^r}{[1-(1-X)^k]}.$$

Collecting terms independent of r into a single constant C gives

$$p_r' = C\,\frac{\Gamma(k+r)}{r!}X^r \quad \text{where} \quad C = \frac{(1-X)^k}{[1-(1-X)^k]}\cdot\frac{1}{\Gamma(k)}.$$

Now, as already remarked (page 94), the parameter k measures (inversely) the variability of λ. A large value of k would be expected if the densities of the various species differed only slightly from one another, and a small value of k would be expected if there were pronounced differences in these densities. In natural communities it is usually found that the differences in abundance among the species is extremely great. This led Fisher (see Fisher, Corbet, and Williams, 1943) to propose that by letting $k \to 0$ in the formula for p_r' an approximation to species-abundance proportions might be obtained.

The limiting form of p'_r as $k \to 0$ is

$$\pi_r = \lim_{k \to 0} p'_r = \gamma \frac{\Gamma(r)}{r!} X^r = \gamma \frac{X^r}{r} \quad \text{for} \quad r = 1, 2, \ldots,$$

where $\gamma = \lim_{k \to 0} C$. One may evaluate γ by noting that

$$\sum_{r=1}^{\infty} \pi_r = -\gamma \ln(1 - X) = 1,$$

whence

$$\gamma = \frac{-1}{\ln(1 - X)}.$$

The expression $\gamma X^r/r$ is the *probability* that a species will be represented in the collection by r individuals. The *expected frequency* of species with r individuals in then

$$n_r = S\gamma \frac{X^r}{r} \equiv \alpha \frac{X^r}{r}, \tag{17.3}$$

where S is the total number of species in the collection and $S\gamma = \alpha$. This is the form in which the logarithmic series is usually give in the ecological literature; it should be emphasized that expressions of the form $\alpha X^r/r$ represent expected frequencies, not probabilities.

We may now express S, the number of species observed, and N, the number of individuals (of all species) in the collection, in terms of the parameters α and X. Thus

$$S = \sum_{r=1}^{\infty} n_r = \sum_{r=1}^{\infty} \frac{\alpha X^r}{r} = -\alpha \ln(1 - X) \tag{17.4}$$

and

$$N = \sum_{r=1}^{\infty} r n_r = \sum_{r=1}^{\infty} \frac{r \alpha X^r}{r} = \frac{\alpha X}{1 - X}. \tag{17.5}$$

These two equations may be solved (see Fisher, Corbet, and Williams, 1943) to give estimates $\hat{\alpha}$ and $\hat{X}$ of the true population values of these parameters, once S and N are known for a particular collection. Estimation of the parameters has also been discussed by Bliss (1965) and by Nelson and David (1967).

The magnitude of X depends only on the size of the sample taken from the parent population. Thus, if we increase the area from which a collection is taken or prolong the duration of operation of a light trap for insects, the only effect will be to change the value of X, *provided* the sample still comes from the same parent population.

The second parameter α is unaffected by sample size and is an intrinsic property of the population. Williams has called it the "index of diversity." It will be seen that, for a given value of n_1, α is proportional to $1/X$. Thus, if α is large, X must be small, and the frequencies $n_2, n_3, \ldots,$ must decrease rapidly. This is equivalent to saying that sparse species are much more common than abundant ones or that the population is made up of a large number of species each with few members; in a word, the population is highly diverse. Conversely, if α is small, X must be comparatively large, the successive frequencies $n_2, n_3, \ldots,$ must decline comparatively gradually, and abundant species are relatively more common.

If a logarithmic series is fitted to an observed frequency table of species-abundances, it is not possible to estimate S^*, the total number of species in the population. To see this note that in arriving at (17.3), namely $n_r = \alpha X^r/r$, we let $k \to 0$ in the negative binomial series; but $k = 0$ implies that the number n_0 of species unrepresented in the collection is infinite. This follows from the fact that [see (17.2)] $p_0 = (1 - X)^k = 1$ when $k = 0$. What was, in fact, assumed was not that $k = 0$ but merely that k was very small indeed. This is tantamount to asserting that the reserve of uncollected species is indefinitely large.

Suppose, now, that a community being studied is enlarged, not by increasing the size of the area sampled, but by widening the range of species to be treated as community members; for example, in dealing with light-trap collections, we could treat only the moths as the community or study the larger community made up of all trapped insects. Similarly, in studying birds in a wooded area, we might consider only warblers or, alternatively, the larger community made up of all birds whatsoever. If the species-abundances in a narrowly defined community formed a logarithmic series, those in a widely defined community might consist of a mixture of several logarithmic series, each with its own parameters. Anscombe (1950) has shown that when this is so one would expect to find larger numbers of very sparse and very abundant species and smaller numbers of species of medium abundance than would be yielded by a single logarithmic series with the same N and S.

3. The Discrete Lognormal Distribution

Consider (17.1) again. To derive the logarithmic series distribution, we substituted the pdf of a Type III distribution for $f(\lambda)$. We shall now let $f(\lambda)$ be the pdf of the lognormal distribution; that is, we assume the λ-values are a sample of size S^* from a distribution having pdf

$$f(\lambda) = \frac{1}{\lambda\sigma\sqrt{2\pi}} \exp\left[-\frac{1}{2\sigma^2}\left(\ln\frac{\lambda}{m}\right)^2\right]. \tag{17.6}$$

Equivalently, $\ln \lambda$ is assumed to be normally distributed with pdf

$$\phi(\ln \lambda) = \frac{1}{\sigma\sqrt{2\pi}} \exp\left[-\frac{1}{2\sigma^2}\left(\ln \frac{\lambda}{m}\right)^2\right].$$

Then $E(\ln \lambda) = \ln m$ and $\text{var}(\ln \lambda) = \sigma^2$. Notice that $\ln m$ is the median as well as the mean of $\ln \lambda$. Therefore m is the median value of λ, or the median abundance.

Substitution of the formula for $f(\lambda)$ in (17.6) into (17.1) gives the discrete lognormal probability

$$p_r = \frac{1}{r!\,\sigma\sqrt{2\pi}} \int_0^\infty e^{-\lambda}\lambda^{r-1} \exp\left[-\frac{1}{2\sigma^2}\left(\ln \frac{\lambda}{m}\right)^2\right] d\lambda$$

$$= \frac{1}{r!\,\sigma\sqrt{2\pi}} \int_0^\infty \exp\left[-\lambda + r \ln \lambda - \frac{1}{2\sigma^2}(\ln \lambda - \ln m)^2\right] \frac{d\lambda}{\lambda}. \quad (17.7)$$

This is the probability that a species in the collection (or sample) will be represented by r individuals, for $r = 0, 1, 2, \ldots$. The probability thus depends on the two parameters σ^2 and m; σ^2 is independent of the size of the sample but m, the median abundance, is a function of sample size.

An alternative form of (17.7), suggested by Grundy (1951), is obtained by putting $\ln \lambda = x$. Then p_r becomes

$$p_r = \frac{1}{r!\,\sigma\sqrt{2\pi}} \int_0^\infty \exp\left[e^{-x} + rx - \frac{1}{2\sigma^2}(x - \ln m)^2\right] dx.$$

There is no explicit expression for this integral and it is not tabulated. Grundy devised a method of evaluating p_r exactly, but Preston (1948), who was the first to test this distribution against field observations, simply used theoretical lognormal frequencies to graduate the observed frequencies, which were, of course, grouped. This is equivalent to supposing that each species is represented by its expected number of individuals and that these numbers are not subject to sampling variation. According to Bliss (1965), such an approximation is quite adequate.

As with the logarithmic series, so also with the discrete lognormal, the observed series is truncated; additional species over and above the S observed ones are assumed to be present in the population and to have escaped collection by chance.

We now consider Preston's (1948) original method of grouping the values of r and, as a consequence, what led him to suggest fitting the lognormal distribution in the first place. Since an observed species-abundance histogram is usually markedly J-shaped (as in Figure 25), with very few high frequencies for low values of r and a long tail representing the few abundant species, it is both natural and convenient to plot r on a logarithmic scale. Moreover, as Preston remarks, "Commonness is a *relative* matter"; one would say that a certain species was so many *times* as common as another. Preston therefore chose grouping into what he calls "octaves" as the most natural procedure; that is, the midpoint of each group is double that of the preceding group.

As group boundaries he takes $r = 1, 2, 4, 8, \ldots$, so that the midpoints of the groups are at $r = 1\frac{1}{2}, 3, 6, 12, \ldots$. A species that falls on a group boundary, for instance one containing 2^x individuals, is treated as contributing half a species to the octave (2^{x-1} to 2^x) and half a species to the octave (2^x to 2^{x+1}). When observations are grouped in this way and plotted (which is equivalent to plotting on semilogarithmic paper using logarithms to base 2) results are obtained of which Figure 26 is typical. The histogram looks as though it

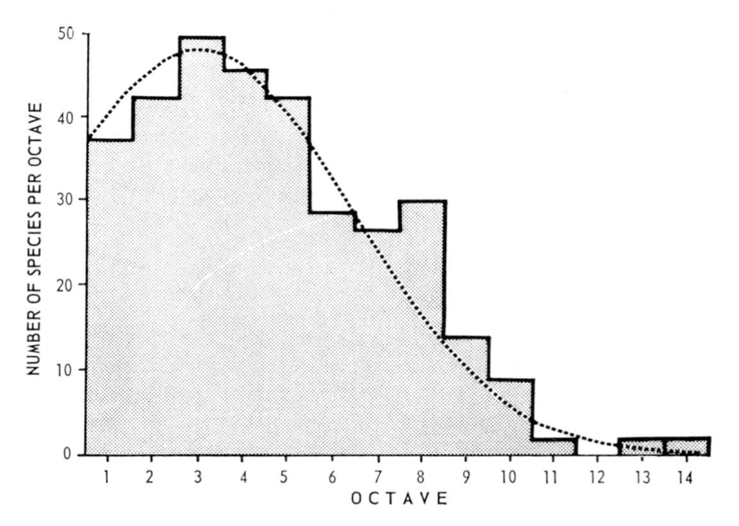

Figure 26. Species abundances in a collection of moths caught in a light trap. Data from Dirks (1937) quoted in Preston (1948). The curve (fitted by Preston) is $n(R) = 48 \exp[-0.207(R - R_0)^2]$, where $n(R)$ is the number of species in the Rth octave and $R_0 = 3$ is the number of the modal octave.

would be well fitted by a symmetrical normal curve truncated on the left. This leads to the idea that the values of λ for the several species might well be lognormally distributed.

Truncation of the curve on the left (at what Preston calls the "veil line") is inevitable. Those species so rare that the expected number of individuals in a sample of the size at hand is less than one are not expected to be found in the sample. Thus the relative abundances of these uncollected species cannot be observed. If the sample size were doubled, the expected number of individuals of every species would be doubled too. The whole curve would then shift one octave to the right or, equivalently, the veil line (or truncation point) would move one octave to the left, and some species that had hitherto been missed would now be found in the collection.

The salient feature of Preston's observations is this: when the species are grouped into octaves, the observed histogram often exhibits a maximum in some octave to the right of the first one. In other words, the observed octave frequencies first increase and then decrease. This could not happen if the species-abundance distribution conformed to a logarithmic series, as we shall now show.

Consider the limiting form of the logarithmic series distribution [see (17.3)], when $X = 1$. This is the harmonic series, α, $\alpha/2$, $\alpha/3$, ... As before, let the observed number of species be S and write F_r for the number of species in the rth octave. Then

$$F_1 = \frac{n_1}{2} + \frac{n_2}{2} \qquad \propto S\alpha(\tfrac{1}{2} + \tfrac{1}{4}) \qquad = 0.750 S\alpha,$$

$$F_2 = \frac{n_2}{2} + n_3 + \frac{n_4}{2} \qquad \propto S\alpha(\tfrac{1}{4} + \tfrac{1}{3} + \tfrac{1}{8}) \qquad = 0.708 S\alpha,$$

$$F_3 = \frac{n_4}{2} + \sum_{i=5}^{7} n_i + \frac{n_8}{2} \quad \propto S\alpha(\tfrac{1}{8} + \tfrac{1}{5} + \tfrac{1}{6} + \tfrac{1}{7} + \tfrac{1}{16}) \qquad = 0.697 S\alpha,$$

$$F_4 = \frac{n_8}{2} + \sum_{i=9}^{15} n_i + \frac{n_{16}}{2} \propto S\alpha(\tfrac{1}{16} + \tfrac{1}{9} + \cdots + \tfrac{1}{15} + \tfrac{1}{32}) = 0.694 S\alpha,$$

....

The frequencies in successive octaves are seen to decrease monotonically. It follows a fortiori that when $X < 1$ the octave frequencies must also decrease monotonically and therefore there cannot be a maximum to the right of the veil line, which is what Preston so often observed. By postulating a lognormal distribution for λ he overcame this shortcoming of the logarithmic series distribution.

When the species abundance curve is lognormal, it becomes possible to estimate S^*, the total number of species in the population including those uncollected. Thus, if the area under the truncated lognormal curve fitted to the data is set equal to S, the number of species in the sample, an estimate of S^* is given by the area under the complete untruncated curve. Bliss (1965) gives an account of the calculations.

4. The Negative Binomial Distribution

As already remarked, to assume that $\ln \lambda$ has a normal distribution ensures that the frequencies of the species-abundances when logarithmically grouped shall increase to some modal value before decreasing, provided the sample is large enough for the veil line to fall to the left of the mode. But the assumption also entails the hypothesis that λ itself has some modal value greater than zero; that is, very rare species as well as very abundant ones are assumed to be less frequent than species having some intermediate value of abundance. It is in this postulate that Preston's hypothesis differs most markedly from that of Fisher, Corbet, and Williams (1943). However, it is not a necessary postulate. Even if $n_1 > n_2 > \cdots > n_r > \cdots$, a plot of species frequencies against octaves may still show a humped distribution.

As an example we assume with Brian (1953) that the observed n_r can be graduated by a truncated negative binomial series for which we do *not* assume that $k \rightarrow 0$. It is then found that a plot of species frequencies against octave numbers may give a humped distribution in the same way that the lognormal series does. In this case, however, the curve is skewed instead of being symmetrical (see Figure 27). As before, unless the sample is large enough, the mode may be to the left of the veil line and therefore hidden. Only with large samples will the humped form of the histogram be manifest.

Suppose, then, that the n_r are assumed to form a negative binomial distribution. This is equivalent to assuming that the densities of the several species (the values of λ) have a Type III distribution. We must now enquire into the shape of this distribution which has pdf (see page 206)

$$f(\lambda) = \frac{1}{\Gamma(k)} P^{-k} \lambda^{k-1} e^{-\lambda/P}.$$

In particular, we wish to know whether $f(\lambda)$ decreases monotonically from its value at $\lambda = 0$ or whether, instead, it has a mode at some $\lambda > 0$.

Assume P to be constant. Then

$$\frac{df(\lambda)}{d\lambda} = \frac{1}{\Gamma(k)} P^{-k} \lambda^{k-2} e^{-\lambda/P} \left(k - 1 - \frac{\lambda}{P} \right).$$

It is seen that, if $k > 1, f(\lambda)$ has a maximum when $\lambda = P(k-1)$, whereas, if $0 \leq k \leq 1$, $df(\lambda)/d\lambda$ is negative for all values of λ.

It follows that if species-abundance data are fitted by a truncated negative binomial series for which $k > 1$ we shall be led to infer that species of intermediate abundance are commoner than rare species, but, if $k \leq 1$, we infer that rare species are the most numerous. Then the *ungrouped* histogram of species frequencies will decrease monotonically with $n_1 > n_2 > \cdots$, whereas

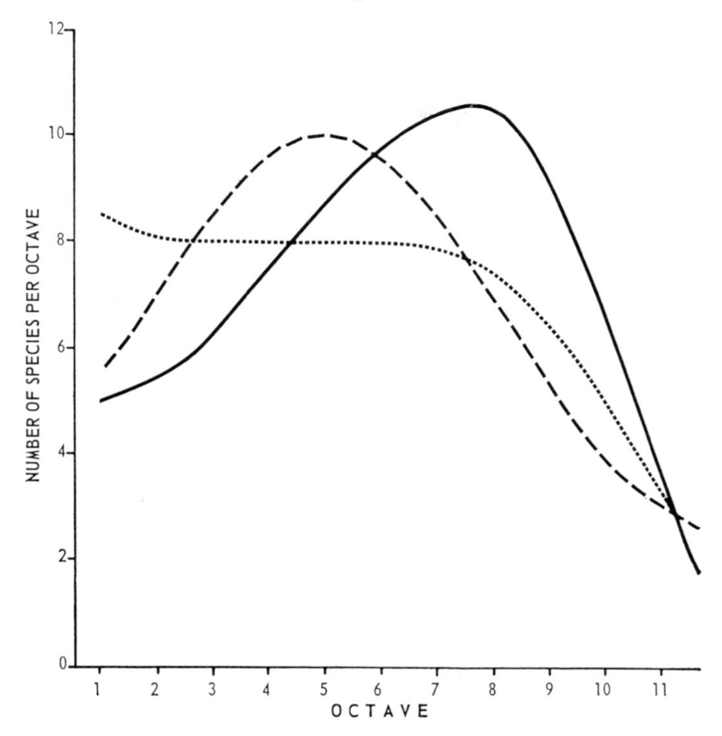

Figure 27. Three theoretical species-abundance curves. Solid line: negative binomial with $k = 0.20$ and $P = 642$; Dashed line: lognormal with $n(R) = 10 \exp[-0.194(R - R_0)^2]$; fitted by Preston (1948). Dotted line: logarithmic series with $\alpha = 11.18$. The curves were fitted to a single set of data on a population of breeding birds. [Data from Saunders, 1936, quoted in Brian, 1953. Redrawn from Brian, 1953].

the histogram of grouped frequencies may have a mode in some octave to the right of the first.

When the negative binomial distribution, with $k > 0$, fits the data, it is possible to estimate S^* from observations on the sample. The probability that a species will contain r individuals is (see page 206)

$$p'_r = \frac{\Gamma(k + r)}{r!\,\Gamma(k)} \frac{P^r}{(1 + P)^{k+r}[1 - (1 + P)^{-k}]}, \; r = 1, 2, \ldots .$$

The mean and variance of this distribution are

$$E(r) = \frac{kP}{1 - (1 + P)^{-k}}$$

and

$$\text{var}(r) = (1 + P + kP)E(r) - [E(r)]^2.$$

The parameters P and k may therefore be estimated from the mean and variance of the empirical distribution. Then, since

$$S^* = \frac{S}{1 - (1 + P)^{-k}},$$

we may estimate S^* from S and the estimates of P and k.

5. The "Broken-Stick" or Ordered Random-Intervals Distribution

Up to this point we have been considering the frequencies n_r of the variate r, the numbers of individuals per species. This approach is impracticable when the population being investigated contains comparatively few species. It is then more reasonable to enquire into the frequencies $N_1, N_2, \ldots,$ and N_s, the numbers of individuals in the S species (see page 203). We assume that every species in the population is represented in the sample so that $S^* = S$. For convenience we rank the species in order of increasing abundance; then N_1 denotes the number of individuals in the rarest species, and N_s the number in the commonest species.

We now show how three different, in fact, contradictory models can lead to the hypothesis that the expected proportion of individuals in the jth species is

$$E\left(\frac{N_j}{N}\right) = \frac{1}{S} \sum_{i=1}^{j} \frac{1}{S + 1 - i}, \tag{17.8}$$

where

$$\sum_{j=1}^{S} N_j = N.$$

All three models seem extremely far-fetched biologically. What is interesting about them, however, is that all lead to the same theoretical distribution of species-abundances.

The Broken-Stick Model

The possibility that species-abundances might be graduated by (17.8) was first suggested by MacArthur (1957), who argued as follows. Suppose S species divide up the environment at random among themselves so that they occupy nonoverlapping niches; also, let the number of individuals in each species be proportional to the size of the species' niche. Then, on the analogy that the niche-space to be divided up may be likened to a line (or "stick") of unit length, we suppose the line to be broken at random into S segments; the lengths of these segments, ranked from smallest to largest, represent the abundances of the 1st, 2nd, . . . , Sth species, ranked from rarest to commonest. The analogy is unsound if it is postulated that what the species divide among themselves is a multidimensional niche-space, since (as we have stressed in Chapters 12 and 16) there is no unique way of randomly partitioning a space of more than one dimension into nonoverlapping parts; the reasoning that applies when a one-dimensional space is randomly broken cannot be extrapolated to spaces of higher dimensionality. However, this objection to the model is easily countered by postulating that what is divided up by the species is some single abundance-limiting factor. If this were so, the broken-stick analogy might hold; the derivation of (17.8) then proceeds as follows. (The argument given here is a paraphrasis of that given by Whitworth (1934) in his Propositions LV and LVI.)

Imagine a line of unit length cut at $S-1$ points located at random on it; then S random segments are formed. Let the lengths of the segments, ranked from smallest to largest, be $l_1, l_2, \ldots, l_S$. Now put

$$l_2 - l_1 = d_1, \quad l_3 - l_2 = d_2, \quad \cdots, \quad l_S - l_{S-1} = d_{S-1};$$

That is, d_r is the difference in length between the rth and $(r+1)$th segments when they are ranked in order of increasing size. Clearly, the length of the original line is given by

$$1 = Sl_1 + (S-1)d_1 + (S-2)d_2 + \cdots + d_{S-1}.$$

Each of the S terms on the right-hand side has equal expectation, since the only condition to which all are subject is that they sum to unity. Therefore

$$E(Sl_1) = E[(S-1)d_1] = E[(S-2)d_2] = \cdots = E(d_{S-1}) = \frac{1}{S},$$

whence

$$E(l_1) = \frac{1}{S^2}, \qquad E(d_1) = \frac{1}{S(S-1)}, \qquad E(d_2) = \frac{1}{S(S-2)},$$

and, in general,

$$E(d_i) = \frac{1}{S(S-i)}.$$

We then see that

$$E(l_2) = E(l_1) + E(d_1) = \frac{1}{S^2} + \frac{1}{S(S-1)},$$

$$E(l_3) = E(l_1) + E(d_1) + E(d_2) = \frac{1}{S^2} + \frac{1}{S(S-1)} + \frac{1}{S(S-2)}.$$

$$\vdots$$

$$E(l_j) = E(l_1) + \sum_{i=1}^{j-1} E(d_i) = \frac{1}{S}\sum_{i=1}^{j} \frac{1}{S+1-i},$$

as in (17.8).

The "Shared Subniches" Model

An entirely different model leading to (17.8) was proposed by Cohen (1966). He calls it the "ordered random intervals" model. Consider the hypothetical multidimensional niche space available to the S species and suppose it to be subdivided into S subniches. The niche of each species consists of a number of these subniches; the first (rarest) species has a niche consisting of one subniche, the niche of the second consists of two subniches, and so on. Then no two species can have exactly coextensive niches, since each of the S species occupies a different number of subniches. But the niches of different species often overlap and sometimes the set of subniches constituting the niche of one species will be a proper subset of the niche of another.

Now consider the jth species which is to occupy j subniches. It is supposed that the number of individuals in this species is proportional to the number of subniches the species would have to "invade" in order to occupy j different subniches; it is assumed that every subniche to be invaded is chosen independently and at random from the S available subniches so that some subniches may be invaded several times over by the jth species. Notice also that each species acts independently of all the others; each subniche is available to all the species.

Cohen (1966) discusses this model in terms of a balls-in-boxes analogy. However, a clearer exposition seems possible in terms of the classic "card collector's problem" (see Feller, 1968). Envisage a deck of cards with S distinguishably different *kinds* of cards present in equal proportions. Cards are drawn from the deck at random (and replaced after each drawing) until j different kinds of cards have been seen. Let N_j be the number of cards drawn up to and including the jth card of a new kind (i.e., of a kind not hitherto seen). Then, as we show below, the expectation of N_j is

$$E(N_j) = S \sum_{i=1}^{j} \frac{1}{S+1-i}. \tag{17.9}$$

The parallel between the card-collecting game and the ecological situation should now be clear: drawing a card is analogous to invading a subniche; drawing the jth new kind of card is analogous to invading the jth hitherto uninvaded (by this species) subniche; once the jth species has invaded j different subniches its niche is complete and it carries out no more invasions.

Formula 17.9 is derived as follows (for convenience, we speak in terms of the card-collecting game). Obviously the first card drawn must be of a kind not hitherto seen so that $N_1 = E(N_1) = 1$. Now denote by X_k the number of cards drawn after the kth new card up to and including the $(k+1)$th new card. Then

$$N_j = 1 + X_1 + X_2 + \cdots + X_{j-1} = \sum_{k=0}^{j-1} X_k, \text{ say,}$$

if we put $X_0 = 1$, and $E(N_j) = \sum_{k=0}^{j-1} E(X_k)$. Now, when k kinds of cards have been seen, there remain $S - k$ kinds of cards still unseen and the probability of drawing one of them, which we may call a success, is therefore $(S - k)/S = p$, say. The probability of a failure, that is of drawing a card of a kind already seen, is $k/S = 1 - p = q$.

Therefore $E(X_k)$ is one more than the expected number of failures preceding the first success in a sequence of Bernoulli trials for which the probability of success is p. The probability of exactly x failures followed by a success is $q^x p$ and the expected number of failures is therefore $\sum_{x=1}^{\infty} xq^x p = q/p$.

Then

$$E(X_k) = 1 + \left(\frac{q}{p}\right) = \frac{S}{S - k}$$

and

$$E(N_j) = \sum_{k=0}^{j-1} E(X_k) = S\left\{\frac{1}{S} + \frac{1}{S-1} + \cdots + \frac{1}{S-j+1}\right\}$$

$$= S \sum_{i=1}^{j} \frac{1}{S+1-i}.$$

We now have

$$E(N_j) \propto \sum_{i=1}^{j} \frac{1}{S+1-i},$$

as in (17.8).

The Model of Complete Independence

Still another hypothesis leading to the distribution (17.8) has been advanced by Cohen (1968). It is simply that the abundances of the S species are proportional to S independently, identically, exponentially distributed random variates. The model contains no reference to niches. The hypothesis leads to the same expected species-abundances as the broken-stick model (for a proof, see Feller, 1966).

Several attempts have been made to fit the broken-stick distribution to field data, with varying success. Cohen (1966) has summarized and discussed many of the results. One would scarcely expect the distribution to be generally applicable, since it has only a single parameter, namely S the number of species. It would be surprising to say the least, if all collections with the same number of species had the same species-abundance proportions.

6. The Species-Area Curve in Vegetation Studies

Most of the work on species-abundance relations has been done by animal ecologists studying large collections of animals. Equivalent studies on plant communities cannot be carried out in the same way, since it is seldom possible to count the numbers of individual plants, of each of the several species, in a sample of vegetation (see page 81). Thus data to which species-abundance distributions may be fitted are generally unobtainable.

An alternative approach consists in collecting data for species-area curves. A tract of vegetation is sampled repeatedly with quadrats of different sizes, and for every quadrat the number of species it contains is recorded. We can then investigate the relationship between quadrat area and the mean number of species per quadrat.

Let us postulate that the vegetation of all species in a given quadrat consists of a number of "plant units." These plant units are not identifiable as such and cannot be counted. Without attempting to define precisely what these units are, we merely assume that their number is proportional to the area of the quadrat. Denote by N the number of plant units in a quadrat of unit area.

Suppose now that we wish to judge whether the species abundances have a logarithmic series distribution. If so, the expected number of species represented by r plant units must be given by a term of the form $\alpha X^r/r$. Writing S_1 for the number of different species in a quadrat of unit area it then follows [from (17.4) and (17.5) on page 207] that

$$N = \frac{\alpha X}{1 - X}$$

and

$$S_1 = -\alpha \ln(1 - X) = \alpha \ln\left(1 + \frac{N}{\alpha}\right). \tag{17.10}$$

Now let the sampling be repeated using quadrats of area q. The number of plant units per quadrat of all species combined becomes Nq and we write S_q for the expected number of species. From (17.10) it is seen that

$$S_q = \alpha \ln\left(1 + \frac{Nq}{\alpha}\right),$$

or, when q is large,

$$S_q \sim \alpha \ln \frac{Nq}{\alpha} = \alpha \ln \frac{N}{\alpha} + \alpha \ln q.$$

It thus appears that if the species-abundance distribution has the form of a logarithmic series, and sampling is done with large enough quadrats, the number of species per quadrat will be proportional to the logarithm of quadrat area.

Hopkins (1955), who sampled several plant communities (including heaths, marshes, woods, and grassland) of area 400 m^2 with quadrats ranging in area up to 100 m^2, found that the observed species versus log area curve was roughly linear for large quadrats, though for small ones for which the expected curve flattens out, the observed numbers of species exceeded expectation (Figure 28 shows an example).

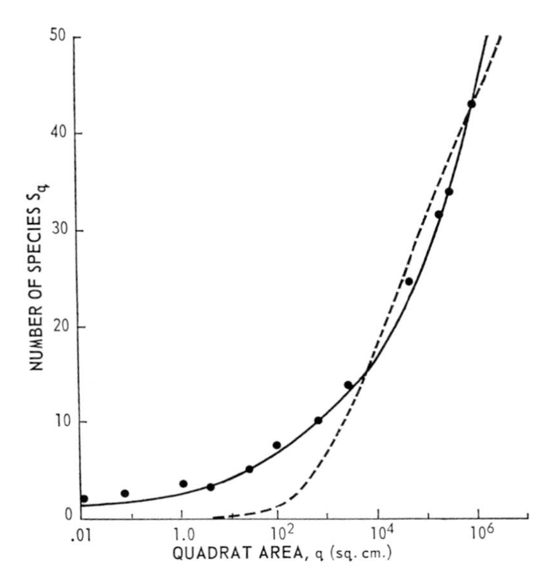

Figure 28. Species-area curves. Dots: data from Hopkin's (1955) Community II. Dashed line: the curve $S_q = 5.7 \ln(1 + q/420)$ fitted by Hopkins. Solid line: the curve $S_q = 2.82\, q^{0.20}$ fitted by Kilburn (1966). (Redrawn from Kilburn (1966) who gives the equation for his curve as $S_q = 17.8\, q^{0.20}$, since he treated 1 m^2 instead of 1 cm^2 as the unit of area.)

Results such as these, however, do *not* permit one to draw conclusions, even approximate ones, about the species-abundance relations of these plants. In fitting the curve $S_q = \alpha \ln(1 + Nq/\alpha)$ to the observations in order to test the hypothesis that the species-abundance frequencies form a logarithmic series, we are tacitly assuming that the vegetation mosaic is so fine-grained that its patches correspond to "plant units" and that the patches of the various species are randomly mingled. Only if both assumptions were true would the contents of a quadrat constitute a random sample from the population. It is clear that to derive theoretical species-area curves we must make assumptions not only regarding the form of the species-abundance curve but also the grain of the vegetation mosaic and the way in which the species are mingled. Very elaborate hypotheses would be needed and to explore their consequences would be difficult.

Attempts to find empirical formulas to describe species-area relations seem worthwhile, however. How such formulas should be interpreted is admittedly not clear; they do not provide evidence bearing directly on the relative abundances of the species, since, as already remarked, the spatial pattern of the vegetation is at least as important as the species-abundance relation in determining the shape of a species-area curve.

Recent empirical studies of species-area curves have been described by Kilburn (1966), who sampled communities of area 900 m^2 with quadrats ranging in size up to 100 m^2; the communities studied were prairie, deciduous forest, and jack-pine woodland. He found that curves of the form $S_q = kq^z$ (with parameters k and z estimated from the data) fitted the observations better than did the curve tested by Hopkins. He inferred that in general it is the logarithm of the number of species rather than the number itself that is proportional to log area. Figure 28 shows examples of the two curves fitted to data from Hopkins (1955); the vegetation was grassland heavily grazed by sheep.

18

Ecological Diversity and Its Measurement

1. Introduction

In Chapter 17 we considered species-abundance curves whose form depends on two things: the number of different species present in a community or population and the relative proportions of their abundances. It is worthwhile to inquire how we might sum up, in two descriptive statistics, the properties of these curves.

When the species-abundance frequencies in an actual collection are well fitted by one or another of the theoretical distributions already described, the parameters of the fitted distribution are obviously suitable as descriptive statistics. If the distribution is lognormal, the appropriate statistics are the estimates of S^*, the total number of species in the population, and σ^2, the variance of the lognormal curve. If the distribution is negative binomial with $k \neq 0$, the appropriate statistics are the estimates of S^* and k (the parameter P depends on sample size). The logarithmic series distribution has two parameters, α and X, but since X is a function of sample size and S^* is assumed to be indefinitely large this leaves only α as a statistic that describes an intrinsic property of the population being sampled. The broken-stick distribution also has only a single descriptive parameter, S, the number of species (assumed to be the same in both sample and population).

The statistics described in the preceding paragraph all suffer from the defect that they are not sufficiently widely applicable. What is needed are descriptive statistics that can be used for any community, no matter what the form of its species-abundance distribution and even when no theoretical series can be found to fit the data.

We begin by considering the properties of any collection, regardless of whether it is to be treated as a population in its own right or as a sample from some larger parent population. Two statistics are clearly needed to describe a collection, of which the first and most obvious is S, the number of species it contains. Now suppose we are dealing with data consisting of a list of the numbers of individuals, $N_1, N_2, \ldots, N_S$, in each of the S species. If the data are portrayed in histogram form, S is the range of the data or the width of the histogram. As a second statistic, to describe the shape of the histogram, we require something analogous to variance. If the N_j were frequencies of some discrete *quantitative* variate, variance as ordinarily calculated would, of course, be the obvious statistic to use, but we are now considering an unordered *qualitative* variate; the individuals are classified according to the species to which they belong and there is no a priori reason for listing them in any particular order. The shape of the histogram is therefore best described in terms of what may be called its "evenness." Thus the distribution has maximum evenness if all the species abundances (the N_j) are equal; and the greater the disparities among the different species abundances, the smaller the evenness.

Before considering evenness per se, however (in Section 5), it is necessary to discuss what has come to be called the "diversity" of a collection.* Various ways of defining and measuring diversity have been proposed and some are discussed here in detail, but it should first be emphasized that diversity, however defined, is a single statistic in which the number of species and the evenness are confounded. A collection is said to have high diversity if it has many species and their abundances are fairly even. Conversely, diversity is low when the species are few and their abundances uneven. It will be seen that since diversity depends on two independent properties of a collection ambiguity is inevitable; thus a collection with few species and high evenness could have the same diversity as another collection with many species and low evenness.

This difficulty did not arise when the notion of diversity was first introduced by Williams (see Fisher, Corbet, and Williams, 1943). On the assumption that most species-abundance distributions would be well fitted by logarithmic series distributions, he proposed that the parameter α of that distribution be used as an index of diversity (see page 208). This index can be applied only if the logarithmic series does indeed fit the species-abundance data, but to determine whether it does may be impossible if there are only a few species and each is represented by a different number of individuals. Thus α is unsuitable as an index of diversity unless the collection at hand has many species and also unless its species abundances form a logarithmic series. Some other measure of diversity is needed. In particular, we require one that will be applicable to small as well as large collections.

* "Diversity" is sometimes used merely as a synonym for "number of species." That is not the sense in which it is used here.

2. Simpson's Measure of Diversity

A useful measure of diversity was proposed by Simpson (1949). (It was mentioned in a different context on page 102.) Suppose two individuals are drawn at random and without replacement from an S-species collection containing N individuals, of which N_j belong to the jth species ($j = 1, 2, \ldots, S$; $\sum_j N_j = N$). If the probability is great that both individuals will belong to the same species, we can say that the diversity of the collection is low. This probability is $\sum_j [N_j(N_j - 1)]/[N(N-1)]$ and so we may use

$$D = 1 - \sum_j \frac{N_j(N_j - 1)}{N(N-1)} \tag{18.1}$$

as a measure of the collection's diversity. When the collection is being treated as a complete population, the value of D so obtained is an exact population parameter, free of sampling error.

Now suppose that the collection is being regarded as a random sample from some indefinitely large parent population with the same number of species. Let p_j denote the proportion of individuals of the jth species in the whole population. The true value of p_j is unknown, since the population has not been fully censused; its maximum likelihood estimator is $\hat{p}_j = N_j/N$. Although we could use $1 - \sum_j \hat{p}_j^2 = 1 - \sum_j (N_j/N)^2$ as an estimator of the population value of D, it is biased. An unbiased estimator is given by

$$\tilde{D} = 1 - \sum_j \frac{N_j(N_j - 1)}{N(N-1)},$$

as we now show (see Good, 1953; Herdan, 1958).

Denote by n_r the number of species represented in the sample by r individuals; that is, n_r is a "frequency of a frequency" (see page 204) and $\sum_r rn_r = N$. Next, put $M = \sum_r r^2 n_r$; it is seen that $M = \sum_j N_j^2$. It remains to prove that $(M - N)/[N(N-1)]$ is an unbiased estimator of $\sum p_j^2 = 1 - D$. Note first that

$$\frac{M - N}{N(N-1)} = \frac{1}{N(N-1)}\left\{\sum_r r^2 n_r - \sum_r rn_r\right\} = \frac{1}{N(N-1)} \sum_r r(r-1)n_r,$$

whence

$$E\left[\frac{M-N}{N(N-1)}\right] = \frac{1}{N(N-1)} \sum_r r(r-1)E(n_r). \tag{18.2}$$

Now, each time an individual is drawn from the population, the probability that it will belong to the jth species is p_j. Therefore the probability that in a sample of size N there will be r representatives of the jth species is

$$\binom{N}{r} p_j^r (1-p_j)^{N-r}.$$

Taking all S species into account, the expected number that will be represented by exactly r individuals is thus

$$E(n_r) = \sum_{j=1}^{S} \binom{N}{r} p_j^r (1-p_j)^{N-r}.$$

Then, from (18.2),

$$E\left[\frac{M-N}{N(N-1)}\right] = \sum_j \sum_r \frac{r(r-1)}{N(N-1)} \cdot \frac{N!}{r!\,(N-r)!} \cdot p_j^r (1-p_j)^{N-r}$$

$$= \sum_j p_j^2 \sum_r \binom{N-2}{r-2} p_j^{r-2} (1-p_j)^{N-r}.$$

The second sum on the right is 1 for all values of j; hence

$$E\left[\frac{M-N}{N(N-1)}\right] = E\left[\frac{1}{N(N-1)} \sum_j N_j(N_j-1)\right] = \sum_j p_j^2.$$

We have therefore shown that an unbiased estimator of population diversity is given by

$$\tilde{D} = 1 - \frac{1}{N(N-1)} \sum_j N_j(N_j-1),$$

since $E(\tilde{D}) = 1 - \sum_j p_j^2$. A comparison of this result with (18.1) shows that the expression for the diversity in a fully censused population is formally identical with that for $\tilde{D}$, the unbiased estimator of the diversity of a population which we have only sampled. The N_j denote population values in the former case and sample values in the latter. However, unlike D, $\tilde{D}$ is subject to sampling variation; its standard error has been derived by Simpson (1949).

3. The Information Measure of Diversity

An entirely different way of measuring diversity may be arrived at as follows (see Khinchin, 1957). It is important to stress at the outset that we are considering here an indefinitely large (effectively infinite) population. The arguments about to be given must be modified when the diversity of a finite population is being measured and a discussion of these modifications is deferred to a later section.

Imagine, then, an infinite population of individuals that can be classified into s classes (or species) $A_1, A_2, \ldots, A_s$. Every individual belongs to one and only one class, and the probability that a randomly selected individual will belong to the class A_j is p_j. Thus $\sum_{j=1}^{s} p_j = 1$. As a measure of the diversity of the population, we wish to find a function of the p_j, $H'(p_1, p_2, \ldots, p_s)$, say, that meets the following conditions:

1. For a given s the function takes its greatest value when $p_j = 1/s$ for all j. Denote this greatest value by $L(s)$. Then

$$L(s) = H'\left(\frac{1}{s}, \frac{1}{s}, \ldots, \frac{1}{s}\right).$$

2. The diversity of the population is unchanged if we postulate the existence of $(s+1)$th, $(s+2)$th, ..., classes to which no individuals belong; that is,

$$H'(p_1, p_2, \ldots, p_s, 0, \ldots, 0) = H'(p_1, p_2, \ldots, p_s).$$

3. Suppose the population is subjected to an additional separate classification process that divides it into t classes, $B_1, B_2, \ldots, B_t$. Each individual belongs to exactly one B-class, and the probability that it will belong to class B_k is q_k with $\sum_{k=1}^{t} q_k = 1$. Then the double classification yields st different classes, $A_j B_k (j = 1, \ldots, s; k = 1, \ldots, t)$, and the probability that a randomly chosen individual will belong to the class $A_j B_k$ may be written π_{jk}. Clearly, if the A-classification and the B-classification are independent, $\pi_{jk} = p_j q_k$, but if the attributes on which the classifications are based are mutually dependent, then $\pi_{jk} = p_j q_{jk}$, where q_{jk} is the conditional probability that an individual will belong to B_k, given that it belongs to A_j. For the diversity of the doubly classified population we may write

$$H'(AB) = H'(\pi_{11}, \pi_{12}, \ldots, \pi_{st}).$$

Next, put $H'_j(B) = H'(q_{j1}, q_{j2}, \ldots, q_{jt})$ for the diversity under the B-classification *within* the class A_j, and put

$$H'_A(B) = \sum_j p_j H'_j(B)$$

for the mean diversity under the B-classification within all the A classes. Condition 3 then is that we should have

$$H'(AB) = H'(A) + H'_A(B).$$

If the classifications are independent so that $q_{jk} = q_k$ for all j, then $H'(AB) = H'(A) + H'(B)$.

Having specified the three conditions that H' is to satisfy, we now show that the only function with these properties is

$$H'(p_1, p_2, \ldots, p_s) = -C \sum_j p_j \log p_j, \tag{18.3}$$

where C is a positive constant.

Before doing so, however, it is desirable to show how the conditions are to be interpreted in an ecological context.

Condition 1 merely ensures that for a population with a given number of species the measure of diversity will be a maximum when all the species are present in equal proportions (or with maximum evenness).

Condition 2 ensures that, given two populations in which the species are evenly represented, the population with the larger number of species will have the higher diversity. This will become clear in the proof.

To understand the implications of condition 3 imagine a forest of s different species of tree and suppose that the trees are also classified according to height. For the purpose of the argument assume height to be a discrete variate; there are t possible heights. The A-classification is the species classification and the B-classification is the height classification. Thus we could measure separately the species diversity of the trees, $H'(A)$, and their height diversity, $H'(B)$. Also, we could calculate their "species-height" diversity, $H'(AB)$, taking both classifications into account. Condition 3 stipulates that if species and height are wholly independent (as in Figure 29*a*) then $H'(AB) = H'(A) + H'(B)$. In this case knowledge of a tree's species gives no information

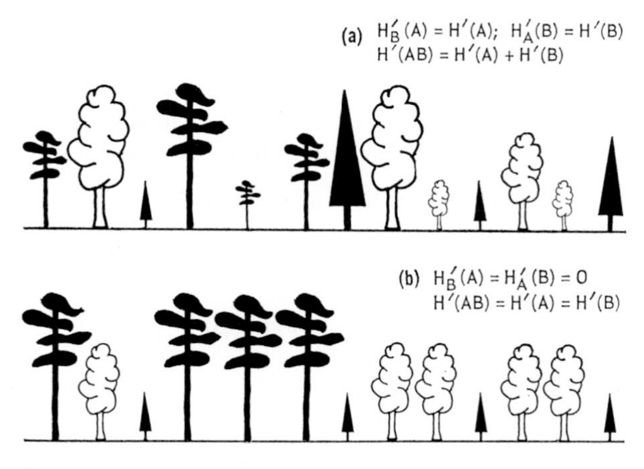

Figure 29.

about its height and vice versa. But, if trees of one species are all of one height, peculiar to the species (or, equivalently, if trees of a given height are all of the same species), as in Figure 29*b*, then $H'_B(A) = H'_A(B) = 0$ and condition 3 now requires that $H'(AB) = H'(A) = H'(B)$. In this case knowledge of a tree's species immediately reveals its height and vice versa; neither classification adds anything to the information yielded by the other.

In deriving (18.3), we begin by dealing with the special case in which $p_j = 1/s$ for all j.

From conditions 2 and 1

$$L(s) = H'\left(\frac{1}{s}, \frac{1}{s}, \ldots, \frac{1}{s}\right) = H'\left(\frac{1}{s}, \frac{1}{s}, \ldots, \frac{1}{s}, 0\right)$$

$$\leq H'\left(\frac{1}{s+1}, \frac{1}{s+1}, \ldots, \frac{1}{s+1}\right)$$

$$= L(s+1).$$

This shows that $L(s)$ is a nondecreasing function of s.

Next, consider m mutually independent classifications of the population, $S_1, S_2, \ldots, S_m$, under each of which the individuals are assigned in equal proportions to r different classes. Then under the kth classification, S_k, for instance, the diversity is $H'(S_k) = L(r)$. This relation holds for all k. Since the classifications are independent, the diversity of the population under the multiple classification is given by

$$H'(S_1 S_2 \ldots S_m) = H'(S_1) + H'(S_2) + \cdots + H'(S_m)$$

$$= m\ L(r).$$

This follows from condition 3. But, since the population has been subdivided into r equal classes at each of m successive classifications, the multiple classification yields r^m ultimate classes, each containing the same proportion of individuals. Therefore

$$H'(S_1 S_2 \ldots S_m) = L(r^m).$$

Thus we see that $m\ L(r) = L(r^m)$.

Likewise, for any other pair of positive integers, say n and s,

$$n\ L(s) = L(s^n).$$

Now, for arbitrary values of r, s, and n let m be chosen so that $r^m \leq s^n \leq r^{m+1}$ or $m \log r \leq n \log s \leq (m+1) \log r$.

Then

$$\frac{m}{n} \leq \frac{\log s}{\log r} \leq \frac{m}{n} + \frac{1}{n}. \tag{18.4}$$

Recall that $L(s)$ is a nondecreasing function of s, as has already been shown. It then follows that

$$L(r^m) \leq L(s^n) \leq L(r^{m+1})$$

or

$$m\ L(r) \leq n\ L(s) \leq (m+1)\ L(r),$$

whence

$$\frac{m}{n} \leq \frac{L(s)}{L(r)} \leq \frac{m}{n} + \frac{1}{n}. \tag{18.5}$$

From (18.4) and (18.5) it now follows that

$$\left| \frac{L(s)}{L(r)} - \frac{\log s}{\log r} \right| \leq \frac{1}{n}.$$

Therefore, since n may be taken as large as we like,

$$\frac{L(s)}{L(r)} = \frac{\log s}{\log r} \quad \text{or} \quad L(s) \propto \log s.$$

Then, writing C for the constant of proportionality, we have

$$H'\left(\frac{1}{s}, \frac{1}{s}, \ldots, \frac{1}{s}\right) = L(s) = C \log s = -C \sum \frac{1}{s} \log \frac{1}{s}$$

as the diversity of an s-species population in which all the species are equally represented.

Now consider the general case in which the p_j are not all equal. As before, we wish to define the diversity of a population classified into s classes, A_1, $A_2, \ldots, A_s$, with a proportion p_j of the individuals in the class A_j. Put $p_j = g_j/g$, where g_j and g are integers. Now perform a second, dependent classification of the population, the B-classification, as follows: B has g classes, all of the same size and labeled $B_1, B_2, \ldots, B_g$. These classes are divided into s groups, with g_j classes in the jth group. If an individual is in the class A_j under the A-classification, it is *defined* as belonging to the jth *group* of B-classes under the B-classification. The probability that it belongs to any given class in this group is the same for all classes in the group and is $1/g_j$.

The double classification process should become clear from an actual example which is shown schematically here. There are three A-classes ($s = 3$) in proportions $p_1 = \frac{6}{10}$, $p_2 = \frac{3}{10}$ and $p_3 = \frac{1}{10}$. There are $g = 10$ B-classes, with $g_1 = 6$ B-classes in group 1 (equivalent to A_1), $g_2 = 3$ B-classes in group 2 (equivalent to A_2), and $g_3 = 1$ B-classes in group 3 (equivalent to A_3).

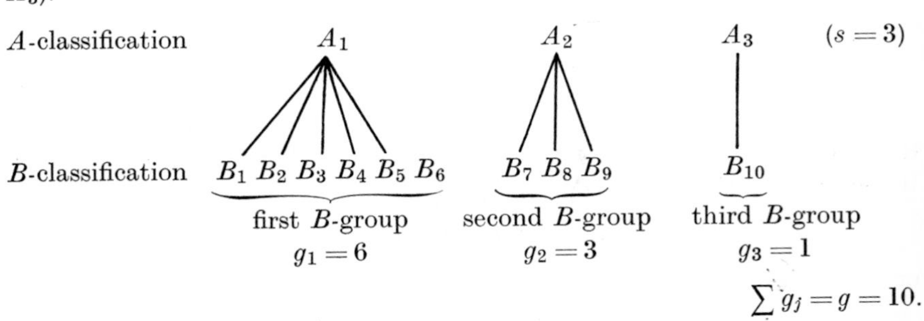

We now see that *within* class A_j the diversity under the B-classification is

$$H'_j(B) = L(g_j) = C \log g_j .$$

It follows that $H'_A(B)$, the *mean* diversity under the B-classification within the A-classes, is

$$H'_A(B) = \sum_j p_j H'_j(B) = C \sum_j p_j \log g_j = C \sum_j p_j \log p_j + C \log g,$$

since $g_j = gp_j$.

Now consider the doubly classified population with classes $A_j B_k$ ($j = 1, \ldots, s$; $k = 1, \ldots, g$). (In the example shown schematically these are the classes in the last line. Giving them their full labels under both classifications, they are $A_1 B_1, \ldots, A_1 B_6, A_2 B_7, \ldots, A_2 B_9$ and $A_3 B_{10}$.) The total number of classes under the double classification is $\sum g_j = g$ and these classes are of equal size. The diversity of the population under the double classification is thus

$$H'(AB) = L(g) = C \log g.$$

Now applying condition 3

$$\begin{aligned} H'(A) &= H'(AB) - H'_A(B) \\ &= C \log g - \left\{ C \sum_j p_j \log p_j + C \log g \right\} \\ &= -C \sum_j p_j \log p_j , \end{aligned}$$

which is what we set out to prove.

This formula is the one proposed by Shannon (see Shannon and Weaver, 1949) as a measure of the information content per symbol of a code composed of s kinds of discrete symbols whose probabilities of occurrence are $p_1, p_2, \ldots, p_s$. In the ecological context H' measures the diversity per individual in a many-species population. It remains to choose the units in which diversity shall be measured. The size of the unit depends on the value assigned to the constant C and on the base used for the logarithms. It is customary to put $C = 1$. Commonly chosen logarithmic bases are the numbers 2, e, and 10. Information theorists use logarithms to base 2 and the information units are then called "binary digits" or "bits." When natural logarithms are used, the unit has been called a "natural bel" (Good, 1950) or a "nat" (McIntosh, 1967b); and with logarithms to base 10 the unit becomes a "bel" (Good, 1950), a "decimal digit" (Good, 1953), or a "decit" (Pielou, 1966a). In ecology the unit to be used and the name to be given it have not yet become standardized.

There has been much debate on whether H' is a suitable measure of ecological diversity. The fact that it measures "information" and "entropy" is beside the point; these fashionable words have been bandied about out of their proper context (the mathematical theory of information) and have led to false analogies that produced no noticeable advance in ecological understanding. It is more illuminating to regard H' as a measure of "uncertainty". If an individual is picked at random from a many-species population, we are uncertain which species it will belong to, and the greater the population's diversity (in an intuitive sense), the greater our uncertainty. The three conditions that H' was required to meet are those appropriate to a measure of uncertainty (Shannon and Weaver, 1949; Khinchin 1957). It is therefore reasonable to equate diversity to uncertainty and use the same measure for both.

Another advantage of using H' is that it permits us to take into account the hierarchical nature of biological classification. Suppose we were comparing two populations and that both had the same number of species in the same relative proportions. Whatever function of these proportions we use as a measure of diversity, the diversities of the two populations must be equal; but, if in one population all the species belonged to a single genus and in the other every species belonged to a different genus, it would be reasonable to regard the latter population as the more diverse of the two. This suggests that it would be desirable to be able to split the diversity measure into two components, a generic component and a specific component. For simplicity we consider only two taxonomic levels, but it is straightforward to extend the argument to as many levels as desired and to split the diversity measure into a corresponding number of components.

Because of the way H' is defined, the splitting is easily done. Let classification of the individuals into their genera be called the G-classification (analogous to the A-classification in the preceding arguments); suppose that there are g genera and that the proportion of individuals belonging to the jth genus is p_j ($j = 1, \ldots, g$). Next, let classification of the individuals into species be called the S-classification (analogous to the previous B-classification); suppose that there are s_j species in the jth genus and that the proportion of individuals belonging to the kth species *within* this genus is q_{jk} ($k = 1, \ldots, s_j$). Then $H'(G) = -\sum_{j=1}^{g} p_j \log p_j$ is the generic diversity and $H'_j(S) = -\sum_{k=1}^{s_j} q_{jk} \log q_{jk}$ is the specific diversity within the jth genus.

The proportion of the individuals in the whole population that belong to the kth species of the jth genus is $p_j q_{jk}$ and therefore the specific diversity of the whole population is

$$\begin{aligned} H'(GS) &= -\sum_{j,k} p_j q_{jk} \log p_j q_{jk} \\ &= -\sum_k q_{jk} \sum_j p_j \log p_j - \sum_{j,k} p_j q_{jk} \log q_{jk} \\ &= -\sum_j p_j \log p_j + \sum_j p_j H'_j(S) \\ &= H'(G) + H'_G(S). \end{aligned} \tag{18.6}$$

(This is merely a restatement of condition 3.)

The terms on the right are the two components sought; $H'_G(S)$ is the specific diversity within a genus averaged over all genera. In two populations with the same numbers of species in the same relative proportions the one with the fewer genera will have the lower $H'(G)$ and the higher $H'_G(S)$.

4. The Information Measure of Diversity in Finite Collections

In Section 3 it was assumed that the population whose diversity was to be measured was indefinitely large. When this is so, the proportions p_j must perforce be estimated from a random sample and we can obtain only an estimate, not the true population value, of H'. Moreover, the estimated H' will be subject to sampling error. We shall not go into the rather complicated subject of estimating H' and its standard error. Details will be found in Miller and Madow (1954), Basharin (1959), Good (1953) and Pielou (1966b, c, 1967a). It suffices to note that the maximum likelihood estimator $-\sum_j (N_j/N) \log (N_j/N)$ (where the N_j are the numbers of members of the jth species in a sample of size N) is biased.

It should also be remarked that $-\sum_j (N_j/N)\log(N_j/N)$ is *not* an appropriate measure of population diversity in a fully censused collection that is being treated as itself a population (Pielou, 1966b) because in information theory H' is strictly defined only for an infinite population; it measures the information content per symbol of a *code* as distinct from that of a *particular message* in the code. We return to this point later.

First it is desirable to consider the following problem: should a fully censused collection be treated as a finite population or as a sample from a larger (conceptually infinite) parent population? A decision on this point is always necessary before we begin to think about diversity. Only when it is unquestionably legitimate to treat a collection as a sample from a larger population is it justifiable to estimate the population value of H' by one of the methods described in the references listed. Often it is not legitimate. Unless we can specify precisely the extent of the postulated parent population and ensure that the collection is a truly random sample from it, it is better to treat a collection as an entity to be studied for its own sake and not as part of something bigger.

For example, when a census is taken of all the nesting birds in a forest, we cannot safely assert that indistinguishable results would have been obtained had adjacent tracts of forest been studied instead; and even if we believe that this is very likely true it would still be necessary to state the exact boundaries of the region regarded as containing the population. Analogous arguments apply to an insect collection caught in a light trap, to a diatom collection in a sample of river water, and to many others. It is obviously absurd to estimate a value of H' that is supposed to pertain to some population unless we can state exactly what that population is.

We therefore require an uncertainty measure appropriate to a finite population. In information theory the information content per symbol of a message made up of N symbols of s different kinds, of which N_j are of the jth kind (see Brillouin, 1962), is

$$H = \frac{1}{N} \log \frac{N!}{N_1! N_2! \cdots N_s!}. \tag{18.7}$$

It will be seen that $N!/(N_1!N_2! \cdots N_s!)$ is the number of distinguishably different permutations (or messages) that could be composed with the given symbols. However, this fact has no ecological relevance.

Now, as already remarked, the analogy between information theory and ecology should not be pushed too far. However, if we accept H' as a good measure of the diversity of a large population, of which we can examine only a sample, it is reasonable to use H for a fully censused collection [as first done by Margalef (1958)]. In other words, if we treat an infinite population as a code, we can treat a particular finite collection as a particular message.

If, in (18.7), all the N_j are very large indeed,

$$H \sim -\sum_j \frac{N_j}{N} \log \frac{N_j}{N}.$$

To see this use natural logarithms and replace the factorials in (18.7) with the approximation $\ln n! = n(\ln n - 1)$.

Then

$$\begin{aligned} H &= \frac{1}{N}\left\{\ln N! - \sum_j \ln N_j!\right\} \\ &\sim \frac{1}{N}\left\{N \ln N - \sum_j N_j \ln N_j\right\} \\ &= -\sum_j \frac{N_j}{N} \ln \frac{N_j}{N}, \end{aligned}$$

which is formally the same as the maximum likelihood estimator of H'. It is not a good approximation to H in practice, for unless *all* the N_j are very large (which seldom happens) the approximation to $\ln n!$ used in deriving it is not sufficiently close. For purposes of evaluating (18.7) it is better to put

$$\ln n! \sim n(\ln n - 1) + \tfrac{1}{2} \ln 2\pi n,$$

which is the commonly used form of Stirling's approximation to the factorial. Alternatively, we may obtain log factorials from tables.

When H is used as a measure of the diversity of a completely censused collection treated as a population, it is, of course, free of sampling error.

We may also split H into components attributable to different taxonomic levels, as done for H' (see page 230). The required formulas are given in Pielou (1967a).

5. The Measurement of Evenness

As remarked on page 222, diversity depends on both the number of species present in a population (or sample) and on their evenness. It is always desirable to treat these two things separately; to state merely that a collection has a certain diversity value is not very informative. As a measure of evenness it is convenient to take the ratio of the observed diversity of a collection to the maximum it could have given the same number of species. This maximum value is attained when the individuals are divided among the species as evenly as possible.

In what follows it is assumed that diversity is measured by H or H' as appropriate.

Suppose, first, that we are dealing with a finite collection (of N individuals belonging to s species) and that its diversity is H. Unless N is an exact multiple of s, the species could not have equal numbers of individuals. Let $N = s[N/s] + r$, where $[N/s]$ is the integer part of N/s. Then a collection with the given N and s would have maximum diversity if $(s - r)$ species contained $[N/s]$ individuals and the remaining r species contained $[N/s] + 1$ individuals. Therefore

$$H_{\max} = \frac{1}{N} \log \frac{N!}{\left\{\left[\frac{N}{s}\right]!\right\}^{s-r}\left\{\left(\left[\frac{N}{s}\right] + 1\right)!\right\}^{r}}.$$

As a measure of the evenness of the collection we may therefore take the dimensionless number $J = H/H_{\max}$.

Next, consider a conceptually infinite population and suppose its diversity H' is known exactly (this could not happen in practice). An s-species population would have maximum diversity if all the species were present in the same proportion, $1/s$, and therefore

$$H'_{\max} = -\sum \frac{1}{s} \log \frac{1}{s} = \log s.$$

The population's evenness is thus given by

$$J' = \frac{H'}{H'_{\max}} = \frac{H'}{\log s}.$$

Notice that, if s is taken as the base of the logarithms in calculating H', $J' = H'$ identically, since $\log_s s = 1$. Thus H' in appropriately chosen units amounts to a measure of evenness.

Estimating the evenness of a population from a sample is not so straightforward. Suppose that we can estimate the population value of H' from a sample and obtain the sampling variance of the estimate; there are various ways of doing this, as mentioned on page 231. Denote the estimate of H' by $\tilde{H}'$. The evenness is then estimated by $\tilde{J}' = \tilde{H}'/\log s^*$ and its sampling variance by $\text{var}(\tilde{J}') = \text{var}(\tilde{H}')/(\log s^*)^2$, where s^* is the number of species in the population (not the sample).

Provided s^* is known, no difficulty arises. Although the number of species in a sample will usually be less than the population number (i.e., $s < s^*$), if the population is of small extent and has been thoroughly explored, s^* is often discovered independently; but if the parent population is large and the sample contains many poorly represented species, it is reasonable to suppose that the population contains other rare species which we are unaware of. Then s^* is completely unknown; only in exceptional circumstances (when the lognormal or negative binomial distribution fits the species-abundance data) can it be estimated from a sample. Therefore it is often impossible to estimate J'; $(\tilde{H}'/\log s)$ inevitably gives an overestimate. This conclusion is obvious on intuitive grounds. A population containing a large number of undiscovered species clearly has far lower evenness than a sample would lead one to believe.

6. McIntosh's Diversity Index

An entirely different way of measuring diversity and evenness has been proposed by McIntosh (1967b). As before, suppose a collection consists of N individuals belonging to s species with N_j individuals in the jth species ($j = 1, \ldots, s$). The collection may be represented by a point in an s-dimensional coordinate frame having coordinates $(N_1, N_2, \ldots, N_s)$. The distance of this point from the origin is $U = \sqrt{\sum_j N_j^2}$. Clearly, for given N, the greater the number of species the smaller U will be. Thus U is a measure of the uniformity of the collection. Its maximum value is attained when the collection contains only one species; then $\max(U) = N$. Its minimum, when every individual belongs to a different species, is $\min(U) = \sqrt{N}$. Since diversity is the complement of uniformity, we may take the N-complement of U as an absolute measure of diversity, say Δ, in a collection of N individuals; that is, $\Delta = N - U$.

As a measure of diversity that is independent of N we may take

$$\frac{\Delta}{\max(\Delta)} = \frac{N - U}{N - \sqrt{N}}.$$

A measure of evenness may be arrived at similarly. When both N and s are given, the uniformity is least when the species are represented as evenly as possible. Assuming that N is an exact multiple of s, there are then N/s individuals in each species and

$$\min(U \mid N, s) = \left[\sum\left(\frac{N}{s}\right)^2\right]^{1/2} = \frac{N}{\sqrt{s}}.$$

The N-complement of $\min(U \mid N, s)$ is thus the maximum value of the diversity for given N and s; that is, $\max(\Delta \mid N, s) = N - N/\sqrt{s}$. Evenness may therefore be measured by

$$\frac{\Delta}{\max(\Delta \mid N, s)} = \frac{N - U}{N - N/\sqrt{s}}.$$

The adjustments to make when N is not an exact multiple of s are obvious.

The fact that these indices proposed by McIntosh are based on the position in an s-dimensional coordinate frame of a point representing the collection is of especial interest. As we shall see in Chapters 19 and 20, the representation of collections by points having as coordinates the amount of each species in each sample is the basis of some methods of ecological classification and "ordination."

19

The Classification of Communities

1. Introduction

Whenever a many-species population—an ecological community—is sampled, the data obtained consist of (a) lists of the species present in each sample unit or (b) records of the amount of each species in each sample unit. Whichever form the data take, that is, whether they are qualitative or quantitative, it is interesting to inquire whether the units are naturally classifiable into distinct groups.

In ecology most of the work on classification has been done by students of vegetation, and the classification problem is considered here in this context. The sample units are quadrats or larger stands of vegetation. When arbitrarily delimited quadrats are used, there is always a risk that the classification obtained may be markedly affected by quadrat size; if stands of vegetation with natural boundaries form the sample units, they may differ greatly in area. These difficulties are merely pointed out here; they will not be discussed further. For convenience we shall speak consistently of the sample units as quadrats, although units much larger than those we should ordinarily describe as quadrats are often used.

A major problem faced by the classifier is whether classification is even appropriate. It is always possible to subdivide a collection of quadrats in one way or another (i.e., classify them), but it does not follow that the vegetation they represent is classifiable into well-defined separate parts. The act of classification does not of itself answer the question: does the vegetation consist of a number of distinct communities or do the communities merge imperceptibly into one another because the vegetation varies continuously? Regardless of the answer to this question, we may still wish to classify as a matter of convenience; an unnatural classification has been called a "dissection" (Kendall and Stuart, 1966). As an analogy, it is often convenient to show the relief of an area in the form of a map with contour layer coloring. Nobody looking at such a map would suppose that the boundaries between colors represented stepwise discontinuities in elevation on the ground.

This brings us to the two well-known theories concerning the nature of vegetation: the *community concept* and the *continuum concept*. A detailed review of these theories has been given by McIntosh (1967a) whose definitions of the two concepts are quoted below. It should first be remarked that probably few ecologists now hold either theory in pure form.

According to the community concept, vegetation is composed of "well-defined, discrete, integrated units which can be combined to form abstract classes or types reflective of natural entities in the 'real world'." If this is true, it is natural to attempt a classification of vegetation. Although transition zones between adjacent communities undoubtedly occur in nature, supporters of the community concept exclude them from consideration on the grounds that they form only a negligible proportion of the total area in any large tract of vegetation.

According to the continuum concept [again quoting McIntosh (1967a)], "vegetation changes continuously and is not differentiated, except arbitrarily, into sociological entities." If this statement describes the true state of affairs, classification is not "natural," although it may still be convenient, for example, when we wish to map vegetation. The fact that obvious clear-cut vegetational discontinuities do occur in nature does not invalidate the continuum concept. A vegetational discontinuity may be due solely to a discontinuity in some abiotic environmental factor (e.g., a sudden change in the bedrock from which the soil is derived) or to a historical accident (e.g., it may be the boundary of an area that was once burned over). The problem therefore is, do abrupt discontinuities often or habitually occur that can*not* be accounted for by extrinsic causes and must therefore be attributed to interactions among the plants themselves?

For any one particular area this question could, at least in principle, be answered as follows. Suppose the area were known to contain no abrupt environmental discontinuities; then, if there were s plant species, each quadrat could be represented by a point in an s-dimensional coordinate frame, any one point having as coordinates the quantity of each species in the quadrat. If the points formed a single hypersphere, we would infer that the vegetation sampled constituted a single homogeneous community. If the points were clustered into two or more separate hyperspheres, it would follow that the vegetation contained as many discrete natural communities as there were clusters of points. If the points formed a single cluster that was not isodiametric, a hyperellipse, say, we would infer that the vegetation was neither homogeneous nor divisible into separate communities, but instead was continuously variable (see Goodall, 1954).

The interpretation of empirical scatter diagrams is not at all straightforward, however. Thus, if the points fell into separate clusters, a disciple of the continuum hypothesis could always argue that a sufficiently diligent search might reveal hitherto undetected environmental discontinuities. Conversely, an apparent hyperellipse might be regarded (by supporters of the community concept) as two hyperspheres linked by intermediate points representing those quadrats in a transition zone that chanced to be unusually wide. Statistical tests may, of course, be done to judge objectively whether a swarm of points should or should not be regarded as forming separate clusters, but they appear to have been used rarely. As Goodall (1966) says, classification techniques explain how classification may be done but not whether it should be done.

If classification is regarded as inappropriate or undesirable, we may still wish to find some method of condensing (and, with luck, clarifying) a mass of field observations. An "ordination" of the data may then be attempted, the aim being to arrange the quadrats in some coordinate frame to display their interrelationships, using as few dimensions as will suffice. Ordination methods are discussed in Chapter 20.

It will be seen from the foregoing that whether to classify or to ordinate is a matter of much controversy. My own opinion is that what seems reasonable, and what we would therefore do, may often depend on the extent of the area whose vegetation is being studied. When the area to be dealt with is small, the paramount property of the vegetation may be its possession of a mosaic pattern; we would then be led to classify. In a large area fine-grained mosaics (if they exist) may be far less conspicuous than gradual trends or gradients; ordination would then be the natural thing to do.

2. The Different Kinds of Classificatory Method

We now proceed to a consideration of methods of classification. Before we can even begin, four choices must be made. We list them before considering each in detail. Should the classification be:

1. Hierarchical or reticulate?
2. Divisive or agglomerative?
3. Monothetic or polythetic?
4. Based on qualitative or quantitative data?

1. In a hierarchical classification the classes at any level are subclasses of classes at a higher level. Ordinary taxonomic classification with such levels as orders, families, genera, and species, is an example. In a reticulate classification the clusters are defined separately and the links between them have the form of a network rather than a tree. Here we consider only hierarchical systems for the reasons given by Williams and Lambert (1966), namely, that they are "better known, less cumbersome and more widely used in ecological work." These three reasons are admittedly not compelling. If

vegetation is truly classifiable, that is, if discrete, nonintergrading communities are the rule rather than the exception, we ought, of course, to determine whether their relationships are reticulate or hierarchical and classify accordingly. To use a hierarchical method is to make untested assumptions about the true relationships among the communities. A hierarchical system is certainly easier to understand and we can only hope that it represents the true state of affairs.

2. In a divisive classification we begin with the whole quadrat collection and divide and redivide it to arrive at the ultimate classes. In an agglomerative classification we start at the bottom and work upward, beginning with the individual quadrats and combining and recombining them to form successively more inclusive groups. Divisive methods have two great advantages: (a) The computations are generally much quicker, since we do not usually continue the subdivision process down to the point at which individual quadrats are recognized as classes. When an agglomerative method is used, we must begin with individual quadrats. (b) Divisive methods are free from the following difficulty that may often arise with agglomerative methods: in the latter the combining process is begun with the smallest units (the quadrats themselves) and these are the ones in which chance anomalies are most likely to obscure the true affinities. The result is that bad combinations may be made at an early stage in the agglomerative process and they will affect all subsequent combinations.

3. In a monothetic classification two "sister" groups are distinguished by the fact that one has and one lacks a single attribute: for instance, the possession of a particular species. In a polythetic classification two groups are combined or separated on the basis of their over-all similarity. Similarity may be measured in many different ways and there are a bewildering number of polythetic methods of classification, but whatever method is used, similarity is always defined so as to depend on a number of attributes rather than only one. A polythetic method thus has the obvious advantage that it can be made to take account of as many properties of the vegetation as we wish to measure or record. A monothetic method is wasteful of information and can lead to meaningless subdivisions if the attribute chosen as distinctive is ecologically unimportant. However, monothetic methods have two great merits. (a) It was remarked above that there are strong reasons for preferring divisive to agglomerative methods. In the detailed discussions of a few classification methods to be given below it is shown that at present divisive-monothetic methods are feasible, whereas for ecological work divisive-polythetic methods are not. (b) In a divisive-monothetic method, in which (as is usually the case) each division is made into quadrats that do and quadrats that do not contain a particular species, the ultimate groups are defined by the set of species each contains. The recognition of these "species-

sets"—groups of co-occurring species—may be a worthwhile objective quite apart from the classifier's primary purpose, namely, classifying the quadrats in a collection.

4. Whether to use qualitative or quantitative data is often decided by circumstances. If some of the species are so ubiquitous that they are present in every quadrat, obviously quantitative data are essential, at least for these species. When we have a choice, the decision hinges on whether small plants, those that do not contribute much mass or volume to the vegetation, are to be treated as important or unimportant to the classification. The use of presence-or-absence data ensures the small species an effect out of all proportion to their quantity, which may be negligibly small. Whether this is desirable or not is a matter for the ecologist's judgment. Many classification methods can be used with either qualitative or quantitative data or, if not, could be modified for use with data of the kind for which they were not originally devised.

It is now apparent that an ecologist must make a number of choices before he can decide on a classification method. At present no strategy exists for making the choices objectively, and the result is that a multitude of methods has been devised and there is continual controversy over their pros and cons. The nature of the problem makes it unlikely that any one method will be voted "best" by a majority of ecologists. Rather than get embroiled in the debate, we consider here four entirely different methods. No attempt has been made to judge their merits. Of the four, one is divisive-monothetic, two are agglomerative-polythetic, and the fourth is divisive-polythetic. The last of these is not, at present, useful in ecological contexts; because of the length of the computations, it is not feasible to use it unless the number of sample units is small (≤ 16). If it were not for this drawback, it could well prove an excellent method, since it combines the merits of both divisive and polythetic methods. It is worth describing for this reason.

3. Association Analysis

This is a divisive-monothetic method (see Williams and Lambert, 1959, 1960). It is applicable when we have for each quadrat a list of the species it contains. The argument underlying the method is as follows. If any pair of species in the vegetation exhibits association, either positive or negative, the vegetation must be heterogeneous, hence classifiable. The quadrats are to be classified by dividing them into groups within which no interspecies associations remain. Then each group can be regarded as homogeneous. To do this we first divide the whole collection on the species that shows the greatest degree of association with other species. Let this species be called the "critical species." Then, if species A, say, is found to be the critical species, the first subdivision of the quadrats is into those containing A, called group (A), and those lacking A, called group (a). Each of the groups, (A) and (a), is now treated as a collection in itself, and the process is repeated until groups are ultimately reached in which there are no significant associations.

To determine the critical species at any stage it is necessary to determine which species in the group of quadrats being subdivided is most strongly associated with the other species in the same group. Williams and Lambert's method of choosing the critical species is to calculate an index of association, say I, between every possible pair of species. Thus, if there are s species in the group, $s(s-1)/2$ indices must be calculated. I is taken as positive, regardless of whether the association is positive or negative. The values of I are entered in an association matrix as follows (the capital letters denote the different species):

	A	B	C	...
A	—	I_{AB}	I_{AC}	
B	I_{AB}	—	I_{BC}	
C	I_{AC}	I_{BC}	—	
...				

The matrix is, of course, symmetrical. I is set equal to zero: (a) on the main diagonal; (b) when the association is indeterminate because all quadrats contain, or all lack, a particular species; (c) when the association fails to reach a previously chosen level of significance; (the 2×2 tables used for judging the association of each species-pair are therefore to be thought of as doubly restricted (see page 162); otherwise the true probabilities are necessarily unknown and the outcomes of the significance tests are not comparable). The elements of the association matrix are then summed by columns (or rows) and the critical species is that having the greatest value of $\sum I$.

It remains to decide how I shall be defined. Williams and Lambert have tried three possible measures of association, namely, X^2, X^2 (corrected), and $\sqrt{X^2/N} = V$; for definitions, see pages 163 and 166.

An analogous method uses quantitative instead of qualitative data (Kershaw, 1961). If we have, not merely presence-or-absence records, but some measure of the amount of each species in each quadrat, the association matrix may be replaced by a covariance matrix.

4. Information Analysis

This method is agglomerative-polythetic (see Williams and Lambert, 1966). It is one of the so-called similarity methods of classification in which small groups that are similar to one another are successively combined into larger groups.

We must therefore devise some means of measuring the similarity between any two groups of quadrats; this is dealt with later. For the present assume that some suitable similarity coefficient has been defined. Beginning with the individual quadrats (which are to be thought of as the initial groups), we determine the similarity coefficient for every pair. Thus, if there are N quadrats, $N(N-1)/2$ similarity coefficients are calculated. The two most similar quadrats are then united to form the first combined group. There are now $N-1$ groups in the collection, $N-2$ of them being single quadrats and the $(N-1)$th a combined pair of quadrats. Next, the similarity coefficients are compared for all the $(N-1)(N-2)/2$ possible pairs of these groups and again the two that are most similar are combined. The process is continued until in the end all the original quadrats have been combined into a single group.

The number of similarity coefficients to be computed is $(N-1)^2$. To see this note that at the first stage $N(N-1)/2$ coefficients must be computed. Once the first pair of quadrats has been combined, the similarities between it and all the remaining groups (single quadrats) have to be computed anew so that $N-2$ new coefficients are needed. After another pair of groups has been combined the similarities of the newly formed group to each of the other $N-3$ groups are computed, and so on. When the process is complete, the number of coefficients that have been obtained is therefore

$$\frac{N(N-1)}{2} + (N-2) + (N-3) + \cdots + 1 = \frac{N(N-1)}{2} + \frac{(N-1)(N-2)}{2}$$
$$= (N-1)^2.$$

We turn now to the problem of judging which two groups in a collection of groups are the most similar. In information-analysis the decision is arrived at as follows. Each group has a certain "information content"; and, if two groups are united, the combined group so formed generally contains more information than either of its component groups. But the more similar the two groups, the smaller the increase in information when the two are combined. Thus at any stage those two groups are to be combined whose union produces the smallest gain in information content.

It remains to consider how the information content of a group of quadrats is to be measured. Suppose the available data are qualitative; that is, for each quadrat we know only whether or not it contains a given species. Let the group consist of n quadrats with a total of s species and let a_j of the quadrats contain the jth species. Then the proportion of quadrats containing this species is a_j/n and the proportion lacking it is $(n-a_j)/n$. Treating the group as made up of n individuals (i.e., quadrats) of two kinds (with and without the jth species) we may then use Shannon's formula (see page 229) to define the information *per quadrat* in respect of the jth species as

$$H'_j = -\frac{a_j}{n}\log\frac{a_j}{n} - \frac{(n-a_j)}{n}\log\frac{(n-a_j)}{n}.$$

(In other words H'_j now measures the "diversity" of the *quadrats.*)

The total information in the group (still with respect to the jth species) is then nH'_j. Summing over all s species, we obtain I, the total information in the group with respect to all the species; that is

$$I = \sum_{j=1}^{s} nH'_j = -\sum_{j=1}^{s}\left[a_j\log\frac{a_j}{n} + (n-a_j)\log\frac{(n-a_j)}{n}\right]$$

$$= sn\log n - \sum_{j=1}^{s}[a_j\log a_j + (n-a_j)\log(n-a_j)].$$

Now consider the first stage of the agglomeration process, when a pair of single quadrats is to be combined into a group for which $n=2$. The information content of a single quadrat is zero by definition. Let two quadrats be combined and let the total number of species in the resulting group be s. Suppose, also, that of these s species t are common to both quadrats and $u=s-t$ were present in only one quadrat. The species may now be labeled so that

$$a_j = 2 \quad \text{for} \quad j = 1, 2, \ldots, t,$$
$$a_j = 1 \quad \text{for} \quad j = t+1, t+2, \ldots, s.$$

Then the information content of the group of $n=2$ quadrats is

$$I = 2s\log 2 - \sum_{j=1}^{t} 2\log 2 = 2(s-t)\log 2 = 2u\log 2.$$

We see that if the two quadrats have identical species lists (i.e., if $u=0$) the group formed by combining them again has zero information content.

Although Williams and Lambert used the information-analysis method of classification only with qualitative data, it is also usable when the data are quantitative. Thus each quadrat may be recorded as containing so many units (e.g., of weight, volume, or cover-area) of species 1, so many of species 2, and so on. The total information content of the quadrat can then be defined either as $\nu H'$ (H' is Shannon's formula for information per unit; see page

229) or, better, as νH (H is Brillouin's formula; see page 232). Here ν is the total number of units, of whatever quantity is measured, of all species taken together. Although, as Williams and Lambert (1966) remark, the statistic I is not defined for continuously varying data, in practice the measured amounts of the species are necessarily discrete; the size of the units depends on the accuracy with which a continuously varying quantity is measured.

5. Orloci's Sums of Squares Method

This is another agglomerative-polythetic method (see Orloci, 1967a).

Assume that we have a collection of n quadrats containing s species altogether. The amount of each species in each quadrat is measured; that is, the data are quantitative. Each quadrat can then be represented by a single point in an s-dimensional coordinate frame. These points, and subsequently the groups formed from them, are to be combined in a sequence of stages. At any stage two groups are to be combined, provided the increase in within-group dispersion resulting from their union is less than it would be had either of the component groups been joined with some other group.

The within-group dispersion Q_n of a group of n points is defined as the sum of squared distances between every point and the group's centroid. The centroid is the point representing the average quadrat of the group and has coordinates

$$\left(\frac{1}{n}\sum_{j=1}^{n} x_{1j}, \frac{1}{n}\sum_{j=1}^{n} x_{2j}, \dots, \frac{1}{n}\sum_{j=1}^{n} x_{sj}\right) = (\bar{x}_1, \bar{x}_2, \dots, \bar{x}_s),$$

where x_{aj} denotes the quantity of the ath species in the jth quadrat and $\bar{x}_a$ is the mean quantity of this species in all quadrats. Then

$$\begin{aligned} Q_n &= \sum_{a=1}^{s}\left[\sum_{j=1}^{n} (x_{aj} - \bar{x}_a)^2\right] \\ &= \sum_{a=1}^{s}\left[\sum_{j=1}^{n} x_{aj}^2 - n\bar{x}_a^2\right] \\ &= \sum_{a=1}^{s}\left[\frac{n-1}{n}\sum_{j=1}^{n} x_{aj}^2 - \frac{2}{n}\sum_{i<j} x_{ai}x_{aj}\right] \\ &= \sum_{a=1}^{s}\left[\frac{1}{n}\sum_{i<j} (x_{ai} - x_{aj})^2\right]. \end{aligned}$$

Here $\sum_{i<j}$ denotes summation over all pairs of points, counting each pair only once. Thus

$$Q_n = \frac{1}{n}\sum_{i<j}\left[\sum_{a=1}^{s} (x_{ai} - x_{aj})^2\right] = \frac{1}{n}\sum_{i<j} d_{ij}^2,$$

where d_{ij} is the distance between the ith and jth points.

We therefore see that the within-group dispersion may be directly calculated from the interpoint distances.

Groups are now combined in pairs according to the following rule. Consider groups u and v, which have within-group dispersions of Q_u and Q_v, respectively, and denote by Q_{uv} the dispersion of the group formed by combining them. The combination of u and v is "permissible" if

$$Q_{uv} - (Q_u + Q_v) < \begin{cases} Q_{uw} - (Q_u + Q_w), \\ Q_{vw} - (Q_v + Q_w), \end{cases}$$

for all w (i.e., for all other groups). Note that groups u and v are combined only if *neither* of them would combine better with some other group. At each stage of the agglomeration *all* permissible pairwise combinations are made before another round of computations begins.

The method just described is based on the absolute distances between pairs of points. Orloci also proposed an alternative method based on "standardized distances." The argument is best illustrated diagramatically (see Figure 30), in which it is assumed that there are only two species.

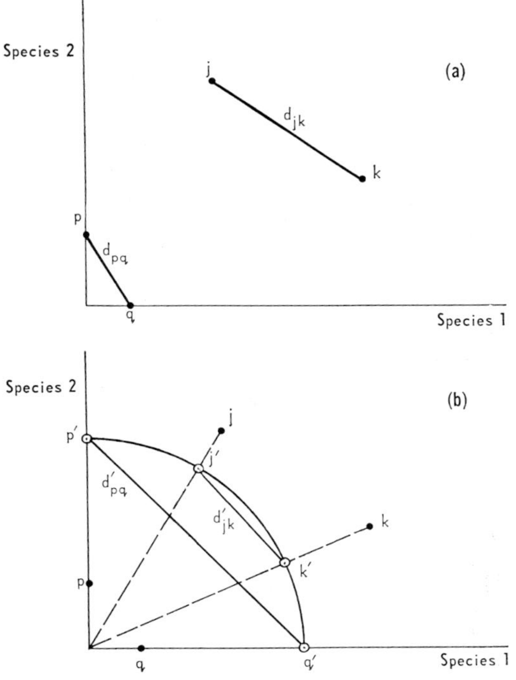

Figure 30. The relation between absolute distances [d_{pq} and d_{jk} in (a)] and standardized distances [d'_{pq} and d'_{jk} in (b)] (After Orloci 1967a).

In Figure 30*a* d_{pq} and d_{jk} show the absolute distances between quadrats p and q and between quadrats j and k. It is seen that $d_{pq} < d_{jk}$; that is, quadrats p and q, which have no species in common but in which the amount of plant material is small, are close together; whereas quadrats j and k, both of which contain large amounts of both species (though in very different proportions), are far apart. If this consequence of using absolute distances is regarded as undesirable, we may use standardized distances instead, as shown in Figure 30*b*. The points representing the quadrats are projected onto a unit circle centered on the origin and the standardized distances between the two pairs of quadrats are then given by d'_{pq} and d'_{jk}, the lengths of the chords $p'q'$ and $j'k'$, respectively. Then $d'_{pq} > d'_{jk}$. The extension to s-dimensions is straightforward and details may be found in Orloci (1967a).

6. The Method of Edwards and Cavalli-Sforza

This is a divisive-polythetic method (see Edwards and Cavalli-Sforza, 1965). As with Orloci's method, so also with this one; each quadrat is represented by a single point in an s-dimensional coordinate frame, and the salient property of a group is the sum of squared distances from the points to the centroid of the group. In the present case the classification process is divisive rather than agglomerative.

Beginning with a swarm of n points, which represent n quadrats, we wish to divide them into two groups in such a way that the within-groups sums of squares will be a minimum and consequently the between-groups sum of squares a maximum. In order to choose which partition has the desired property all possible partitions must be tested. If there are n quadrats in the collection, they can be divided into two groups in $2^{n-1} - 1$ different ways. To see this assume n is even; note that there are $\binom{n}{1}$ ways of dividing the n points into groups of sizes $n-1$ and 1; $\binom{n}{2}$ ways of dividing them into groups of sizes $n-2$ and 2, ..., and $\binom{n}{n/2}$ ways of dividing them into two equal groups of size $n/2$. The total number of possibilities is thus

$$\binom{n}{1} + \binom{n}{2} + \cdots + \binom{n}{n/2}$$

$$= \frac{1}{2}\left\{\left[\binom{n}{0} + \binom{n}{1} + \cdots + \binom{n}{n-1} + \binom{n}{n}\right] - \binom{n}{0} - \binom{n}{n}\right\}$$

$$= \frac{1}{2}[(1+1)^n - 2] = 2^{n-1} - 1.$$

Similar arguments yield the same result when n is odd.

For every possible partition we now perform what is in effect a one-way analysis of variance in which the variates are vectors in s dimensions. Let

Q = the sum of squared distances from all points to the centroid of the whole swarm.

Also, with $i = 1, 2$, let

q_i = the sum of squared distances from the points in group i to the group's own centroid.

Q_i = the squared distance from the centroid of group i to the centroid of the whole swarm.

Then, if there are n_i points in the ith group,

$$Q = \sum_{i=1}^{2} q_i + \sum_{i=1}^{2} n_i Q_i .$$

As already proved (page 244), Q, q_i, and Q_i may all be found by summing the squares of the interpoint distances (taking each distance only once) and dividing by the number of points concerned. We may therefore tabulate these squared interpoint distances in the form of a half-matrix:

Points	α	β	γ	$\cdots$
α		$d^2_{\alpha\beta}$	$d^2_{\alpha\gamma}$	$\cdots$
β			$d^2_{\beta\gamma}$	$\cdots$
γ				
$\vdots$				

Then $Q = (1/n) \sum d^2$, where the summation is over all elements in the half-matrix and there are n points.

Now consider how a particular partition is to be tested. For concreteness assume that $n = 7$ and let the quadrats be labeled $\alpha, \beta, \ldots, \eta$. Let the partition to be tested be that into groups $(\alpha, \beta, \gamma, \delta)$ and (ε, ξ, η) so that $n_1 = 4$ and $n_2 = 3$.

Then

$$q_1 = \frac{1}{n_1}(d^2_{\alpha\beta} + d^2_{\alpha\gamma} + \cdots + d^2_{\gamma\delta})$$

and

$$q_2 = \frac{1}{n_2}(d^2_{\varepsilon\xi} + d^2_{\varepsilon\eta} + d^2_{\xi\eta}).$$

The within-groups sum of squares sought is $q_1 + q_2$. To decide which of the $2^{7-1} - 1 = 63$ possible partitions of the seven quadrats is to be made we must calculate all 63 values of $(q_1 + q_2)$ and choose that partition for which $(q_1 + q_2)$ is smallest. In this manner the original collection of quadrats is divided and the resultant groups redivided until we obtain a classification as fine as desired.

The computing time for this method may be enormously long. Thus at the first stage $2^{n-1} - 1$ partitions have to be tested. Suppose the chosen partition yields groups with n_1 and $n_2 = n - n_1$ points, respectively. Then at the second stage there are $2^{n_1-1} + 2^{n_2-1} - 2$ partitions to test and so on. According to Edwards and Cavalli-Sforza (1965), an initial collection of "16 points can be treated in a reasonable time." For 41 points, however, Gower (1967) states that using a computer with 5 μsec access time the process would require more than 54,000 years. Some form of short cut seems called for, perhaps along the lines suggested by Dagnelie (1966), who comments on the possibility of ruling out inadmissible partitions. As an example of an inadmissible partition, consider a swarm of four points in a plane; if one of them were inside a triangle formed by the other three, it would be inadmissible to treat the interior point as one group and those at the vertices of the surrounding triangle as the other.

7. Hierarchical Levels

Classification of a quadrat collection by any of the methods described, or indeed by any hierarchical method, enables us to construct a tree diagram, or dendrogram, to show the sequence in which the divisions or unions of the groups were made. The tree thus has a number of nodes (shown on the tree by horizontal lines), each of which represents an intermediate group that was divided into two subgroups (in a divisive classification) or formed by the union of two subgroups (in an agglomerative classification). These intermediate groups are of different status; some may be highly heterogeneous and others more homogeneous. The way in which the groups differ in heterogeneity, or in hierarchical level, may be shown by the heights of the nodes in the tree diagram.

For example, consider the following diagrams which show two possible sets of relationships among five ultimate groups, A, B, ..., E; these may be individual quadrats or homogeneous unsplittable groups.

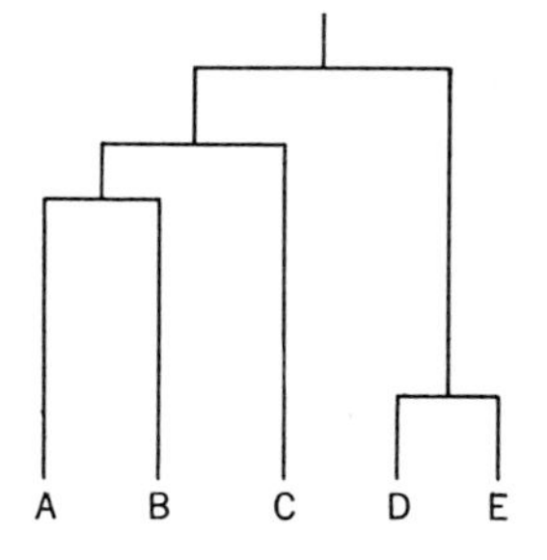

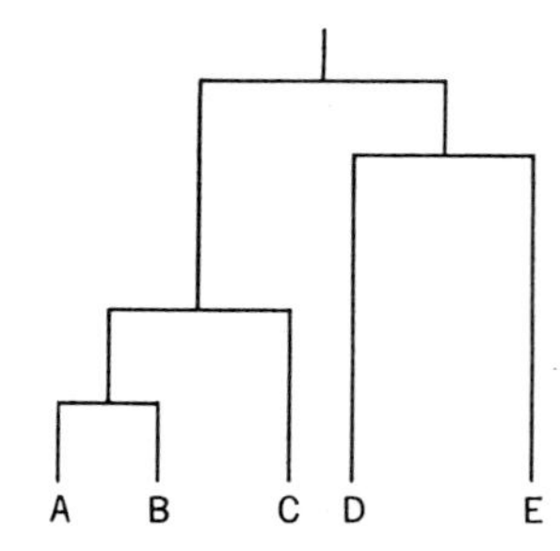

In both trees the constituent members of the intermediate groups are the same, but their hierarchical levels differ. Thus in the tree on the left group (AB) is far more heterogeneous than group (DE), whereas in the tree on the right the opposite is true. Several other differences are visible on inspection.

To portray the different hierarchical levels in a tree diagram we must therefore have some means of measuring the heterogeneity within any group. Then, when the smallest groups (or the individual quadrats) are arrayed along a horizontal base line, the heterogeneity of any intermediate group is shown by the height of the corresponding node above the base.

Various measures of heterogeneity have been used. When classification is by the method of association-analysis (page 240), Williams and Lambert (1960) propose use of the highest individual X^2 value in the association table for the group. This measure is suggested on the grounds of expedience; there is little to recommend it theoretically. When the classification is by information-analysis (page 242), the best measure of intragroup heterogeneity is the total information content of a group (Williams and Lambert, 1966). When the within-group sums of squares is used as the classification criterion, a measure of a group's heterogeneity is given by its mean within-group dispersion (Orloci, 1967a).

8. Concluding Remarks

There is now a huge literature on ecological classification; here we have described in detail only four of the many different methods that have been proposed. Each method consists of a set of rules for operations to be carried out on the raw data. The vexing problem of the method that is best must be left open; it is not even possible to define "best" in this context. As Williams and Lambert (1966) have said, "the difficulty ... is to find objective criteria in an essentially subjective situation." All that we can reasonably demand of a method is that "the major groupings which arise shall not be fewer than, or markedly different from, those recognized intuitively as distinct ecological entities ... ." Work is continuing in the hope that formal classifications will not always be done merely to determine whether the results accord with the investigator's preconceptions. The ultimate aim is to find an intrinsically sensible objective method that gives consistently good results in all circumstances.

20

The Ordination of Continuously Varying Communities

1. The Purpose of Ordination

As already remarked (page 244), we can, conceptually, represent the data obtained by measuring the amounts of s species in each of n sample units by a scatter diagram of n points in an s-dimensional coordinate frame. Classification consists in subdividing the swarm of points into a number of disjoint sets or groups in what we believe is a natural manner. If the points chance to fall into several compact, widely separated groups no difficulty arises and formal rules for effecting a classification are scarcely needed. This ideal result is hardly ever obtained when vegetation is sampled. More often than not the points (representing the quadrats) are diffusely scattered and any classification procedure is largely arbitrary.

A way out of this difficulty is to ordinate the quadrats rather than to classify them. The purpose, as in classification, is still to simplify and condense the mass of raw data yielded by vegetation sampling in the hope that relationships among the plant species and between them and the environmental variables will be manifested. Ordination consists in plotting the n points in a space of fewer than s dimensions in such a way that none of the important features of the original s-dimensional pattern is lost. Ideally we would plot them in two or three dimensions so that they might be easily visualized. As a method of summarizing the results of a survey, ordination has two great advantages over classification; it obviates the need for setting up arbitrary criteria for defining the classes and there is no need to assume that distinct classes (if there are any) are hierarchically related.

2. Principal Component Analysis

Many possible methods of ordination have been devised. The most straightforward is to project the original s-space onto a space of fewer dimensions in such a way that the arrangement of the points suffers the least possible distortion. To take the simplest conceivable case, suppose we wished to project a swarm of points in the plane onto a line to obtain a linear ordination. The distortion will be a minimum if the line is oriented so as to preserve as far as possible the spacing of the points. An artificial example is given in Figure 31*a* which shows results that might be obtained by sampling vegetation made up of only two species of plants. The points represent the quantities

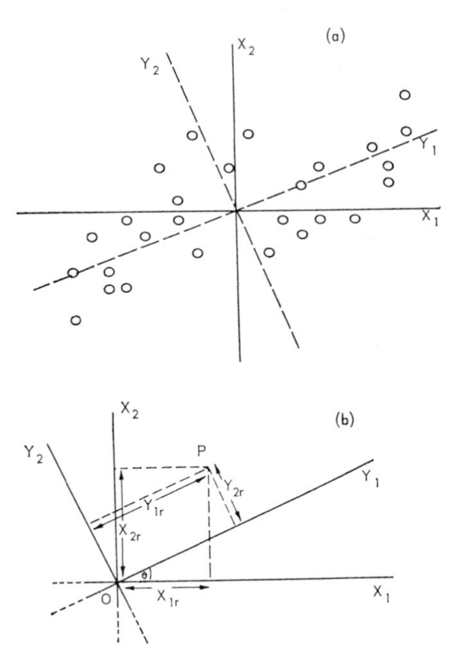

Figure 31*a*. The principal component axes y_1 and y_2 for the swarm of points whose original coordinates were given as (x_1, x_2). Figure 31*b*. To illustrate the rotation of coordinate axes (see text).

of species 1 (measured along the x_1-axis) and of species 2 (on the x_2-axis) in each of $n = 27$ quadrats. The origin of the coordinates is at $(\bar{x}_1, \bar{x}_2)$, the mean quantities of the species averaged over all quadrats.

We now wish to subject the axes to a rigid rotation through an angle θ, say, to form new orthogonal (perpendicular) axes shown in the figure as the y_1- and y_2-axes. Then the y_1-axis is the line $x_1/\cos\theta = x_2/\cos\phi$, where $\phi = \pi/2 - \theta$. (We write $\cos\phi$ in place of $\sin\theta$ to emphasize that $\cos\theta$ and $\cos\phi$ are the direction cosines of the line.) A point with coordinates (x_{1r}, x_{2r})

relative to the old axes now has coordinates (y_{1r}, y_{2r}) relative to the new axes. A linear ordination of the points may be obtained by projecting them onto the y_1-axis, and we may say that the distortion produced by the ordination is least when the sum of squares of the y_1 values, $\sum_{r=1}^{n} y_{1r}^2$, is a maximum. Our object now is to determine what value of θ will give this result.

Notice first that

$$\sum_1^n x_{1r}^2 + \sum_1^n x_{2r}^2 = \sum_1^n y_{1r}^2 + \sum_1^n y_{2r}^2.$$

That this is so may be seen from Figure 31*b* in which P is the point (x_{1r}, x_{2r}) in the original coordinates and (y_{1r}, y_{2r}) in the new coordinates. Clearly $(OP)^2 = x_{1r}^2 + x_{2r}^2 = y_{1r}^2 + y_{2r}^2$. It follows also that since the origin is at the centroid of the swarm of points (in other words, since $\bar{x}_1 = \bar{x}_2 = \bar{y}_1 = \bar{y}_2 = 0$)

$$\text{var}(x_1) + \text{var}(x_2) = \text{var}(y_1) + \text{var}(y_2). \tag{20.1}$$

Figure 31*b* also shows that

$$y_{1r} = x_{1r} \cos\theta + x_{2r} \sin\theta \quad \text{and} \quad y_{2r} = -x_{1r} \sin\theta + x_{2r} \cos\theta.$$

This may be written in matrix form as

$$\begin{pmatrix} y_{1r} \\ y_{2r} \end{pmatrix} = \begin{pmatrix} \cos\theta & \sin\theta \\ -\sin\theta & \cos\theta \end{pmatrix} \begin{pmatrix} x_{1r} \\ x_{2r} \end{pmatrix}, \qquad r = 1, 2, \ldots, n,$$

or, more compactly, as

$$\mathbf{Y} = \mathbf{UX}, \tag{20.2}$$

where

$$\mathbf{U} = \begin{pmatrix} \cos\theta & \sin\theta \\ -\sin\theta & \cos\theta \end{pmatrix}$$

and $\mathbf{X}$ is the $(2 \times n)$ data matrix

$$\begin{pmatrix} x_{11} & x_{12} & \cdots & x_{1n} \\ x_{21} & x_{22} & \cdots & x_{2n} \end{pmatrix};$$

$\mathbf{Y}$ is defined similarly. It is seen that $\mathbf{U}$ is orthogonal; that is, $\mathbf{UU}' = \mathbf{U}'\mathbf{U} = \mathbf{I}$.

We also see that the covariance matrix of the x's is

$$\mathbf{\Sigma}_x = \frac{1}{n}\mathbf{XX}' = \frac{1}{n}\begin{pmatrix} \sum_1^n x_{1r}^2 & \sum_1^n x_{1r}x_{2r} \\ \sum_1^n x_{1r}x_{2r} & \sum_1^n x_{2r}^2 \end{pmatrix} = \begin{pmatrix} \text{var}(x_1) & \text{cov}(x_1, x_2) \\ \text{cov}(x_1, x_2) & \text{var}(x_2) \end{pmatrix}. \tag{20.3}$$

We return now to the problem of determining the value of θ that will make $\sum y_{1r}^2 = n\,\text{var}(y_1)$ a maximum. It is clear from (20.1) that maximizing $\text{var}(y_1)$ is equivalent to minimizing $\text{var}(y_2)$.

From (20.2) we have

$$\mathbf{YY}' = \mathbf{UXX}'\mathbf{U}' \tag{20.4}$$

or

$$\begin{pmatrix} \text{var}(y_1) & \text{cov}(y_1, y_2) \\ \text{cov}(y_1, y_2) & \text{var}(y_2) \end{pmatrix} = \begin{pmatrix} \cos\theta & \sin\theta \\ -\sin\theta & \cos\theta \end{pmatrix} \begin{pmatrix} x_1 \\ x_2 \end{pmatrix} (x_1 x_2) \begin{pmatrix} \cos\theta & -\sin\theta \\ \sin\theta & \cos\theta \end{pmatrix}, \tag{20.5}$$

where for conciseness we have replaced the $(2 \times n)$ data matrix with the vector $(x_1\ x_2)'$.

From (20 5)

$$\text{var}(y_1) = x_1{}^2 \cos^2\theta + x_1 x_2 \sin 2\theta + x_2{}^2 \sin^2\theta.$$

Obviously $\text{var}(y_1)$ is maximized when

$$\frac{d}{d\theta}\,\text{var}(y_1) = 0;$$

that is, when

$$(-x_1{}^2 + x_2{}^2)\cos\theta \sin\theta + x_1 x_2 \cos 2\theta = 0.$$

From (20.5) it is also seen that

$$\begin{aligned} \text{cov}(y_1, y_2) &= (-x_1{}^2 + x_2{}^2)\cos\theta\sin\theta + x_1 x_2(\cos^2\theta - \sin^2\theta) \\ &= (-x_1{}^2 + x_2{}^2)\cos\theta\sin\theta + x_1 x_2 \cos 2\theta. \end{aligned}$$

Thus $\text{var}(y_1)$ is a maximum [and $\text{var}(y_2)$ a minimum] when $\text{cov}(y_1, y_2) = 0$; the new variates y_1 and y_2 are uncorrelated.

Therefore, when $\text{var}(y_1)$ is a maximum, (20.4) becomes

$$\mathbf{U\Sigma}_x\mathbf{U}' = \begin{pmatrix} \lambda_1 & 0 \\ 0 & \lambda_2 \end{pmatrix} = \mathbf{\Lambda} \quad \text{or} \quad \mathbf{U\Sigma}_x = \mathbf{\Lambda U}, \tag{20.6}$$

where $\lambda_i = \text{var}(y_i)$ for $i = 1, 2$.

It is now clear that λ_1 and λ_2 are the latent roots of the symmetric matrix $\mathbf{\Sigma}_x$; they are the roots of the determinantal equation $|\mathbf{\Sigma}_x - \lambda\mathbf{I}| = 0$, and the rows of $\mathbf{U}$ are the latent vectors of $\mathbf{\Sigma}_x$.

Therefore θ may be found by solving the equation

$$\text{var}(x_1)\cos\theta + \text{cov}(x_1, x_2)\sin\theta = \lambda_1 \cos\theta.$$

We are now in a position to do the following. Instead of defining any point by the coordinates (x_1, x_2), we can treat it as the point (y_1, y_2), where

$$y_1 = x_1 \cos\theta + x_2 \sin\theta, \qquad y_2 = -x_1 \sin\theta + x_2 \cos\theta.$$

Each of the two new variates is thus a linear combination of the original variates (the measured quantities of the species). The new variates are so defined that y_1, which is known as the first principal component, has maximum possible variance. The best linear ordination of the points is obtained by projecting them onto the y_1-axis, the first principal axis. In this simple two-dimensional case the direction of the y_2-axis (the second principal axis) is given once that of the first is known, since it must be orthogonal to it. We have also shown above that the two new variates y_1 and y_2 are uncorrelated.

Now consider the general s-dimensional case. Our objective is to find the rigid rotation of the original axes, or equivalently the linear combination of the original variate values (the x's), that will yield derived variates (the y's) with the following properties:

(1). The variance of the y_1's is to be as great as possible.

(2). The variance of the y_2's is to be as great as possible, subject to the restriction that the y_2-axis must be orthogonal to the y_1-axis. The variates y_1 and y_2 are uncorrelated.

(3). The variance of the y_3's is to be as great as possible, subject to the restriction that the y_3-axis must be orthogonal to the y_1- and y_2-axes. There are no correlations among the variates.

. . . .

(s) The final axis, the y_s-axis, is to be orthogonal to all the $(s-1)$ axes already fixed.

This process is principal component analysis.

What we wish to find are the s direction cosines of each of the s principal axes, which are given by the elements of the $s \times s$ matrix $\mathbf{U} = \{u_{ij}\}$ with $i, j = 1, \ldots, s$. They are obtained by solving the matrix equation

$$\mathbf{U}\mathbf{\Sigma}_x\mathbf{U}' = \mathbf{\Lambda} \qquad \text{[this is the same as (20.6)]}.$$

Here $\mathbf{\Sigma}_x$ is the $s \times s$ covariance matrix of the x's and $\mathbf{\Lambda}$ is the diagonal matrix whose elements (the latent roots of $\mathbf{\Sigma}_x$) are $\lambda_j = \text{var}(y_j)$ $(j = 1, \ldots, s)$. The direction cosines of the jth principal axis are the elements of the jth row of $\mathbf{U}$ (which is the jth latent vector of $\mathbf{\Sigma}_x$).

In terms of the original coordinates, the y_j-axis (the jth principal axis) is the line

$$\frac{x_1}{u_{j1}} = \frac{x_2}{u_{j2}} = \cdots = \frac{x_s}{u_{js}}.$$

The jth derived variate (the jth principal component) is

$$y_j = u_{j1}x_1 + u_{j2}x_2 + \cdots + u_{js}x_s.$$

Thus the coordinates of the rth point $(r = 1, \ldots, n)$ are given by

$$y_{jr} = u_{j1}x_{1r} + u_{j2}x_{2r} + \cdots + u_{js}x_{sr} \qquad (j = 1, \ldots, s).$$

These assertions about the general s-dimensional case are not proved here. Proofs may be found in books on multivariate analysis; for example, Kendall (1957), Anderson (1958), Kendall and Stuart (1966), and Morrison (1967).

We have ranked the transformed variates so that

$$\text{var}(y_1) > \text{var}(y_2) > \cdots > \text{var}(y_s).$$

Then, depending on how much of the information in the original data we are willing to sacrifice, we may disregard the variates $y_{k+1}, y_{k+2}, \ldots, y_s$ (those with smallest variances) for some chosen k and, retaining only the k variates with largest variances, ordinate the data in a space of k dimensions. It is, in fact, often found that the first few latent roots of $\mathbf{\Sigma}_x$ account for a large proportion of the total variance; for example, Orloci (1966), analyzing data on the vegetation of sand dunes and dune slacks and considering the 101 most frequent species, found that the first three principal components accounted for more than 40% of the total variance. Likewise, Greig-Smith, Austin, and Whitmore (1967) present a three-dimensional ordination of forest types in the British Solomon Islands Protectorate. In both the examples quoted it was found that clusters of points recognizable in two- or three-dimensional plots of the vegetation samples could be associated with environmental differences.

3. Practical Considerations in Principal Component Analysis

When we wish to ordinate vegetation by means of a principal component analysis, there are four decisions to make.

1. A method of measuring the amount of each species in each sample unit must be chosen. If the sample units are quadrats, there are several ways of measuring the amount of each species; for instance, by counting individuals or by determining fresh weight, dry weight, basal area, or cover. When larger sample units are used—whole plots or stands of vegetation—a convenient method (Orloci, 1966; Gittins, 1965) is to sample each stand with a number of quadrats and take as a measure of the quantity of a species in the stand the frequency of that species in the quadrats; that is, the number of quadrats that contained the species.

2. Whether to standardize the original variates must be decided. If they are standardized the covariance matrix $\mathbf{\Sigma}_x$ in (20.6) is replaced by a correlation matrix but the analysis is otherwise unaltered. Standardization is commonly done when the x variates are measured in several different units: for instance, in psychological studies when measurements on people may be a mixture of test scores, ages, incomes, and other dissimilar quantities measured on entirely different scales. In botanical work, in which the amounts of all the

species in a sample are measured by the same method and in the same units, standardization is not necessary. If the raw data are standardized before analysis, those species whose abundances vary only slightly (usually the less abundant species) will have a greater influence than if unstandardized variates are used. Whether this is desirable is debatable. It should be noticed that standardizing the original data and then finding the principal components does not lead to the same result as finding principal components first and then standardizing the derived variates.

3. The number of principal components to extract—or of axes to use in the final ordination—must be chosen. If the number is to be small, a compromise is necessary between simplicity and precision; this is a matter of subjective judgment. In this connection we must also consider the following. Up to this point we have treated the observations as a population and not as a sample, subject to error, from some larger parent population. If the observations are to be treated as a sample, what we have obtained are the *sample* principal components. It may well happen that even though the s roots of (20.6) (when the elements of $\mathbf{\Sigma}_x$ are sample values) are all different, so that the s principal components of the sample can be found unambiguously, the population values of the $s-t$ smallest roots are equal. If this is so, there is no point in extracting more than t components; the remaining $s-t$ can have any direction as long as they are orthogonal to one another and to the previously extracted components. Methods exist for testing whether the $s-t$ smallest sample roots could have come from a parent population in which these roots are equal (see Kendall, 1957; Lawley and Maxwell, 1963; and references therein). These tests require that the parent distribution be s-variate normal, which is probably seldom true in the context we are considering—that of plant species on units of ground. Transformation of the data to normalize it is necessary before any test can be performed. In any case, for the purpose of ordinating vegetation we usually wish to use comparatively few components, those we can confidently regard as "real."

4. An entirely different form of component analysis, known as Q-type analysis, is sometimes performed. (The usual analysis described in the preceding pages is called R-type analysis.) Formally, the two analyses are identical, but whereas in R-type analysis we began by obtaining the $s \times s$ matrix $\mathbf{\Sigma}_x = (1/n)\mathbf{X}\mathbf{X}'$ [see (20.3)], a Q-type analysis starts from the $n \times n$ matrix $(1/s)\mathbf{X}'\mathbf{X}$. The rth diagonal element of this matrix, $(1/s)\sum_{j=1}^{s} x_{jr}^2$, is the variance of the species quantities (all species) within the rth quadrat; the (r, t)th element is the covariance of the species quantities in the rth and tth quadrats. We are, in effect, contemplating an n-dimensional scatter diagram in which there are s points that represent species. Instead of treating the quantity of a species as an attribute of a quadrat, we treat quantity in a quadrat as an attribute of a species. The result is an ordination of species,

not quadrats. Its interpretation in ecological contexts is not so intuitively clear as the usual R-type quadrat ordination. Moreover, as Sokal and Sneath (1963) have remarked, "the Q-matrix ... does not have a sampling distribution expected of ordinary correlations." This is because the species within a quadrat are (presumably) not independent, whereas in ordinary R-type analysis we generally assume that the observed quadrats are independent. If $n < s$, that is, if there are fewer quadrats than species, the Q-type matrix is of lower order than the R-type and less unwieldy from a computational point of view.

We can, in any case, perform an R-type analysis by means of a "*Q-technique*" (Gower, 1966; Orloci, 1967b). This is possible, since a swarm of n points occupies a space of $n - 1$ dimensions at most, even when they are plotted in an s-dimensional frame with $s > n$. It follows that, regardless of whether we are doing an R- or Q-type analysis, we can perform the computations with an $n \times n$ or an $s \times s$ matrix. Usually we choose the smaller.

As on page 252, let $\mathbf{X}$ denote the $s \times n$ data matrix;

Now write

$$\mathbf{R} = \mathbf{X}\mathbf{X}' \qquad (\mathbf{R} \text{ is an } s \times s \text{ matrix})$$

and

$$\mathbf{Q} = \mathbf{X}'\mathbf{X} \qquad (\mathbf{Q} \text{ is an } n \times n \text{ matrix}).$$

and suppose that $s > n$.

Let α be a latent root of $\mathbf{R}$ and $\mathbf{a}$, the corresponding latent vector so that

$$\mathbf{R}\mathbf{a} = \alpha\mathbf{a} \quad \text{or} \quad \mathbf{X}\mathbf{X}'\mathbf{a} = \alpha\mathbf{a}. \tag{20.7}$$

Similarly, let β and $\mathbf{b}$ be a root and vector of $\mathbf{Q}$ so that

$$\mathbf{Q}\mathbf{b} = \beta\mathbf{b} \quad \text{or} \quad \mathbf{X}'\mathbf{X}\mathbf{b} = \beta\mathbf{b}. \tag{20.8}$$

Premultiplying (20.7) by $\mathbf{X}'$ gives

$$\mathbf{X}'\mathbf{X}(\mathbf{X}'\mathbf{a}) = \alpha(\mathbf{X}'\mathbf{a}) \quad \text{or} \quad \mathbf{Q}(\mathbf{X}'\mathbf{a}) = \alpha(\mathbf{X}'\mathbf{a}).$$

Comparing this with (20.8), we see that $\alpha = \beta$ and $\mathbf{X}'\mathbf{a} = \mathbf{b}$. Thus the latent roots and vectors of $\mathbf{R}$ (the larger matrix) are directly obtainable from those of $\mathbf{Q}$ (the smaller matrix).

4. Principal Coordinate Analysis

In principal component analysis every quadrat is represented by a point whose coordinates are the quantities of each of the s species in that quadrat.

As a result, the distance d_{ij} between the points representing quadrats i and j is

$$d_{ij} = \left[\sum_{t=1}^{s} (x_{ti} - x_{tj})^2\right]^{1/2},$$

where x_{ti} and x_{tj} denote the quantities of species t in quadrats i and j.

In effect, we treat the Euclidean distance between the plotted points as a measure of the difference between the two quadrats. However, this distance is not necessarily the best measure of a *difference*. If we were given only two quadrats, the ith and jth, say, and were asked to propose a suitable measure of the difference between them, a number of possibilities would spring to mind; for example, we might choose to define the difference m_{ij}, say, as

$$m_{ij} = \sum_{t=1}^{s} |x_{ti} - x_{tj}|,$$

that is, as the sum of the absolute magnitudes of the difference for each species. This is only one among several acceptable definitions for m_{ij}. Another possibility is considered in Section 5.

Having chosen a suitable way of measuring interquadrat difference, the question now is, is it possible to plot points (representing the quadrats) in such a way that the distance between each pair of points is equal to this difference? Gower (1966, 1967b) has shown that it is possible and describes a way of doing it. He calls the process principal coordinate analysis.

What we wish to find are the coordinates of the n points; these coordinates are now *not* equal to the quantities of the species in the quadrats.

Denote by $\mathbf{C}$ the $s \times n$ matrix

$$\mathbf{C} = \{c_{tj}\}, \qquad (t = 1, \ldots, s; j = 1, \ldots, n).$$

The elements of the jth column of $\mathbf{C}$, namely $(c_{1j}\ c_{2j}\ \cdots\ c_{sj})'$ are the coordinates of the jth point; these are the values we shall now find.

The squared distance between the ith and jth points is

$$\sum_{t=1}^{s} (c_{ti} - c_{tj})^2 = \sum_t c_{ti}^2 + \sum_t c_{tj}^2 - 2\sum_t c_{ti} c_{tj}. \tag{20.9}$$

This distance² is to be *defined* as m_{ij}^2, the square of the difference between the ith and jth quadrats measured in whatever way we choose. These differences are obtained by observation and constitute the data.

Now put

$$\mathbf{C}'\mathbf{C} = \mathbf{A}.$$

Thus $\mathbf{A}$ is the symmetric $n \times n$ matrix

$$\mathbf{A} = \begin{pmatrix} \sum_t c_{t1}^2 & \sum_t c_{t1}c_{t2} & \cdots & \sum_t c_{t1}c_{tn} \\ \sum_t c_{t2}c_{t1} & \sum_t c_{t2}^2 & \cdots & \sum_t c_{t2}c_{tn} \\ \cdots & \cdots & \cdots & \cdots \\ \sum_t c_{tn}c_{t1} & \sum_t c_{tn}c_{t2} & \cdots & \sum_t c_{tn}^2 \end{pmatrix}$$

$$\equiv \{a_{ij}\}, \text{ say,}$$

where $a_{ij} = \sum_t c_{ti} c_{tj}$. Then, from (20.9)

$$m_{ij}^2 = m_{ji}^2 = a_{ii} + a_{jj} - 2a_{ij}; \qquad m_{ii}^2 = 0. \tag{20.10}$$

The known values m_{ij}^2 $(i, j = 1, \ldots, n)$ may be written as elements of an $n \times n$ symmetric matrix $\mathbf{M}$. From the given $\mathbf{M}$ we must find $\mathbf{A}$ and then $\mathbf{C}$ whose columns give the coordinates sought. The problem is solved as follows:

Note first that we want the origin of the coordinates to be at the centroid. Therefore we must have $\sum_{j=1}^{n} c_{tj} = 0$ for all t. Then the sum of the elements of the ith row of $\mathbf{C}'\mathbf{C} = \mathbf{A}$ is

$$\sum_{j=1}^{n} a_{ij} = \sum_{j=1}^{n} \sum_{t=1}^{s} c_{ti} c_{tj} = \sum_{t=1}^{s} c_{ti} \left(\sum_{j=1}^{n} c_{tj} \right) = 0; \tag{20.11}$$

that is, all the row sums of $\mathbf{A}$ are zero; as also are the column sums, since $\mathbf{A}$ is symmetric.

It will be found that the requirements of (20.10) and (20.11) are fulfilled if we put

$$a_{ij} = -\frac{1}{2} m_{ij}^2 + \frac{1}{2n} \sum_{i=1}^{n} m_{ij}^2 + \frac{1}{2n} \sum_{j=1}^{n} m_{ij}^2 - \frac{1}{2n^2} \sum_i \sum_j m_{ij}^2 . \tag{20.12}$$

Using this formula, $\mathbf{A}$ may be found from $\mathbf{M}$. Now observe that since $\mathbf{A}$ is symmetric we can find an orthogonal matrix $\mathbf{V}$ such that $\mathbf{A} = \mathbf{V\Lambda V}'$, where $\mathbf{\Lambda}$ is the diagonal matrix whose elements are the latent roots of $\mathbf{A}$, that is, the roots of $|\mathbf{A} - \lambda\mathbf{I}| = 0$.

Now we have that $\mathbf{A} = \mathbf{C}'\mathbf{C} = (\mathbf{V\Lambda}^{1/2})(\mathbf{\Lambda}^{1/2}\mathbf{V}')$.

Assume that $s < n$ and recall that $\mathbf{C}'$ and $\mathbf{C}$ have dimensions $n \times s$ and $s \times n$, respectively. Convert them to $n \times n$ matrices $\mathbf{\Gamma}'$ and $\mathbf{\Gamma}$ by adding $n - s$ columns of zeros on the right side of $\mathbf{C}'$ (and, equivalently, $n - s$ rows of zeros at the bottom of $\mathbf{C}$). Then

$$\mathbf{C}'\mathbf{C} = \mathbf{\Gamma}'\mathbf{\Gamma} = (\mathbf{V\Lambda}^{1/2})(\mathbf{\Lambda}^{1/2}\mathbf{V}')$$

so that

$$\mathbf{\Gamma} = \mathbf{\Lambda}^{1/2}\mathbf{V}'.$$

The jth column of $\boldsymbol{\Gamma}$ is $(c_{1j}\ c_{2j}\ \cdots\ c_{sj}\ 0\ \cdots\ 0)'$; these are the coordinates of the jth point in n-space. The jth column of $\mathbf{C}$ is $(c_{1j}\ c_{2j}\ \cdots\ c_{sj})'$, which are the coordinates of the points in s-space. Gower (1966) also shows that these coordinates are referred to principal axes; that is, the axes are oriented to meet the requirements listed on page 254.

To demonstrate the process here is a simple numerical example with $s=2$ and $n=3$ (see Figure 32).

Consider three points, X, Y, and Z, in the x_1-x_2 plane. Their coordinates are $(-7, 1)$, $(8, 2)$ and $(-1, -3)$, respectively. The centroid is at the origin. The distances between the points are

$$d_{xy}^2 = 15^2 + 1^2, \qquad d_{xz}^2 = 6^2 + 4^2, \qquad d_{yz}^2 = 9^2 + 5^2,$$
$$d_{xy} = 15.03, \qquad d_{xz} = 7.21, \qquad d_{yz} = 10.30.$$

As a measure of the *differences* between the pairs of points, take the sums of the distances (all treated as positive) parallel with the two coordinate axes. Then

$$m_{xy} = 15 + 1 = 16; \qquad m_{xz} = 6 + 4 = 10; \qquad m_{yz} = 9 + 5 = 14.$$

The matrix $\mathbf{M}$ with elements m^2 is thus

$$\mathbf{M} = \begin{pmatrix} 0 & 256 & 100 \\ 256 & 0 & 196 \\ 100 & 196 & 0 \end{pmatrix}.$$

The elements of $\mathbf{A}$ are found by substituting the elements of $\mathbf{M}$ in (20.12). It will be found that

$$\mathbf{A} = \frac{4}{3}\begin{pmatrix} 43 & -41 & -2 \\ -41 & 67 & -26 \\ -2 & -26 & 28 \end{pmatrix}.$$

Neglecting the factor $\frac{4}{3}$, we find the latent roots of $\mathbf{A}$ from

$$|\mathbf{A} - \lambda\mathbf{I}| = \lambda(\lambda^2 - 138\lambda + 3600) = 0,$$

whence

$$\lambda_1 = 103.07345; \qquad \lambda_2 = 34.92655; \qquad \lambda_3 = 0.$$

Solving $\mathbf{AV} = \mathbf{V\Lambda}$ yields

$$\mathbf{V} = \begin{pmatrix} 0.53788 & 0.61429 & 0 \\ -0.80093 & 0.15867 & 0 \\ 0.26305 & -0.77296 & 0 \end{pmatrix}.$$

The columns of $\mathbf{V}$, which have been normalized so that $\sum_j v_{ij}^2 = 1$, are the latent vectors of $\mathbf{A}$. Notice that $\sum_j v_{ij} = 0$; this is a necessary consequence of

(20.11). Now put

$$\mathbf{\Gamma} = \mathbf{\Lambda}^{1/2}\mathbf{V}' = \begin{pmatrix} 5.46 & -8.13 & 2.67 \\ 3.63 & 0.94 & -4.57 \\ 0 & 0 & 0 \end{pmatrix}.$$

The first two elements of the columns of $\mathbf{\Gamma}$ are the coordinates of the new points spaced the desired distances apart. These distances are

$$d'_{xy} = \{[5.46 - (-8.13)]^2 + [3.63 - 0.94]^2\}^{1/2} = 13.85,$$
$$d'_{xz} = \{[5.46 - 2.67]^2 + [3.63 - (-4.57)]^2\}^{1/2} = 8.66,$$

and

$$d'_{yz} = \{[-8.13 - 2.67]^2 + [0.94 - (-4.57)]^2\}^{1/2} = 12.12,$$

and it is seen that

$$13.85 : 8.66 : 12.12 = 16 : 10 : 14 = m_{xy} : m_{xz} : m_{yz}.$$

The new points are shown in the lower graph of Figure 32. The direction of the c_1-axis has been reversed (the values increase from right to left) to make comparison of the graphs easier.

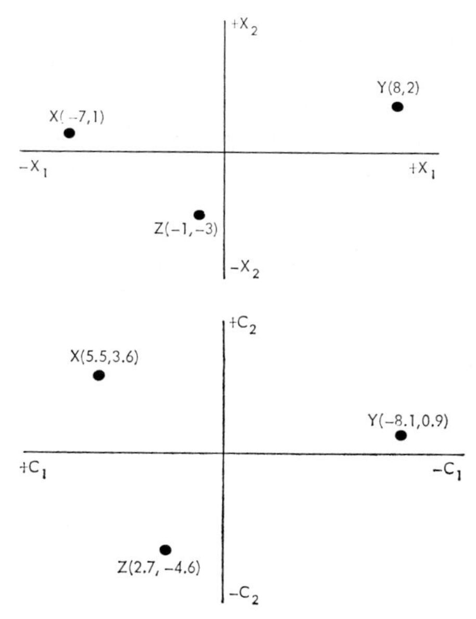

Figure 32. The upper graph shows the quantities of species 1 (x_1-axis) and species 2 (x_2-axis) in three quadrats, X, Y, and Z. The lower graph shows the result of a principal coordinate analysis (the c_2-axis is reversed).

5. Principal Coordinate Analysis with Qualitative Data

Suppose now that only presence-and-absence data have been recorded; for each quadrat we have merely a list of the species it contains and no observations of the amounts of each species. Now score 1 for the presence of a species and 0 for its absence; then, if the quadrats are represented by points in an s-dimensional frame, using the scores of the species as coordinates, all the points will be concentrated at the vertices of an s-dimensional hypercube. A principal component analysis of the points would be formally possible but not illuminating.

We can, however, use as a measure of the *difference* between any two quadrats the number of species occurring in one or the other but not both; for example, if one quadrat contained species A, B, C, and D and another contained C, D, and E, the difference between them would be 3, since the three species A, B, and E occur only once in the pair of quadrats. By a principal coordinate analysis we may plot points, representing the quadrats, so that between every pair the distance is equal to the difference measured in this way.

It is interesting to note that the difference measured, as just described, is equal to the square of the distance between the points when they are plotted at the vertices of a hypercube. This may be shown by an example. Consider two quadrats obtained from vegetation with eight plant species, $A, \ldots, H$. Denoting the presence of a species by a capital letter and its absence by the corresponding lower-case letter, let the quadrats be $(ABcDeFgh)$ and $(ABCdEfgh)$. If these quadrats were represented in 8-space by points having species-scores as coordinates, the coordinates of the points would be $(1, 1, 0, 1, 0, 1, 0, 0)$ and $(1, 1, 1, 0, 1, 0, 0, 0)$. Obviously the square of the distance between these points is 4, which is the number of species that occur in only one of the quadrats.

21

Canonical Variate Analysis

1. Introduction

Chapter 20 dealt with a way of simplifying and condensing the raw data obtained from a vegetation survey. Each quadrat (or other sample unit) yielded a vector of variate values (the amounts of the several species), and it was shown how, by the technique of principal component analysis, we can find a few linear combinations of the variates that account for nearly all the variance in the data.

However, observations on the quantities of the different species in a unit of vegetation may constitute only one of the sets of observations an ecologist makes. He may, in addition, measure several environmental variates; for example, the water-holding capacity of the soil, the quantity of soil organic material, the slope of the ground, the depth of the soil, and various other measurable properties. These environmental data constitute a second set of observations, qualitatively different from the first set. We are naturally led to inquire into the interrelationships among the two *sets* of variates. If the amounts of only one plant species had been recorded and measurements made on only one feature of the environment, it would be straightforward to measure the correlation between them. What we now require is an analogous method appropriate to multivariate data. Canonical variate analysis is such a method.

As in principal component analysis, we are concerned with linear combinations of the observations that have certain desirable properties. Before explaining how these combinations are arrived at, we shall describe the "input" and "output" of a canonical variate analysis and consider what the analysis achieves. A preview of the output (in symbolic form) will show what is to be gained by examining the relationships among linear combinations of the original variates rather than among the raw data and will provide a motive for the matrix manipulations needed to attain the result.

The input to the analysis is as follows: suppose p environmental variates, measured on each sample unit, yield a vector of observations $(x_1\ x_2\ \ldots\ x_p)$ and suppose the amounts of q species yield the vector $(y_{p+1}\ y_{p+2}\ \cdots\ y_{p+q})$. It is convenient to number the species from $p+1$ to $p+q$ rather than from 1 to q. Also, it is assumed that $p \leq q$; if the number of species is less than the number of measured environmental factors, we need only reverse the labeling of the two sets of data.

Now write $(\mathbf{x}'\ \mathbf{y}')$ for the $(p+q)$-element row vector representing all the observations on any one unit. The $(p+q) \times (p+q)$ covariance matrix of all the data is

$$\mathbf{\Sigma} = \begin{pmatrix} \mathbf{x} \\ \mathbf{y} \end{pmatrix} (\mathbf{x}'\mathbf{y}') = \left(\begin{array}{ccc|ccc} \sigma_{11} & \cdots & \sigma_{1p} & \sigma_{1,\,p+1} & \cdots & \sigma_{1,\,p+q} \\ \cdots & \cdots & \cdots & \cdots & \cdots & \cdots \\ \sigma_{p1} & \cdots & \sigma_{pp} & \sigma_{p,\,p+1} & \cdots & \sigma_{p,\,p+q} \\ \hline \sigma_{p+1,\,1} & \cdots & \sigma_{p+1,\,p} & \sigma_{p+1,\,p+1} & \cdots & \sigma_{p+1,\,p+q} \\ \cdots & \cdots & \cdots & \cdots & \cdots & \cdots \\ \sigma_{p+q,\,1} & \cdots & \sigma_{p+q,\,p} & \sigma_{p+q,\,p+1} & \cdots & \sigma_{p+q,\,p+q} \end{array}\right)$$

$$= \begin{pmatrix} \mathbf{\Sigma}_{11} & \mathbf{\Sigma}_{12} \\ \mathbf{\Sigma}_{21} & \mathbf{\Sigma}_{22} \end{pmatrix}. \qquad (21.1)$$

We have partitioned $\mathbf{\Sigma}$ into four parts. The $p \times p$ submatrix $\mathbf{\Sigma}_{11}$ is the covariance matrix of the x's (the environmental variates); similarly, the $q \times q$ submatrix $\mathbf{\Sigma}_{22}$ is the covariance matrix of the y's (the species quantities). And the elements of the $p \times q$ submatrix $\mathbf{\Sigma}_{12}$ denote covariances of the form $\text{cov}(x, y)$. Thus $\sigma_{i,\,p+j}$ (with $i = 1, \ldots, p; j = 1, \ldots, q$) is $\text{cov}(x_i, y_{p+j})$ or the covariance of the ith environmental variate and the quantity of the $(p+j)$th species. Since $\mathbf{\Sigma}$ is symmetric, we see that $\mathbf{\Sigma}_{21}$ is simply the transpose of $\mathbf{\Sigma}_{12}$ or $\mathbf{\Sigma}_{21} = \mathbf{\Sigma}'_{12}$.

Our object now is to find p linear combinations of the x's, say,

$$\xi_i = a_{i1}x_1 + a_{i2}x_2 + \cdots + a_{ip}x_p, \qquad (i = 1, \ldots, p),$$

and q linear combinations of the y's, say,

$$\eta_j = b_{j1}y_{p+1} + b_{j2}y_{p+2} + \cdots + b_{jq}y_{p+q}, \qquad (j = 1, \ldots, q),$$

such that the $(p+q) \times (p+q)$ correlation matrix of the derived variates (the ξ's and η's) has the form

$$
\mathbf{P} = \begin{pmatrix} \boldsymbol{\xi} \\ \boldsymbol{\eta} \end{pmatrix} (\boldsymbol{\xi}' \boldsymbol{\eta}') = \left(\begin{array}{cccc:cccc:c}
1 & 0 & \cdots & 0 & \rho_1 & 0 & \cdots & 0 & \\
0 & 1 & \cdots & 0 & 0 & \rho_2 & \cdots & 0 & \mathbf{0}_{p\times(q-p)} \\
\cdots & \cdots & \cdots & \cdots & \cdots & \cdots & \cdots & \cdots & \\
0 & 0 & \cdots & 1 & 0 & 0 & \cdots & \rho_p & \\
\hdashline
\rho_1 & 0 & \cdots & 0 & 1 & 0 & \cdots & 0 & \\
0 & \rho_2 & \cdots & 0 & 0 & 1 & \cdots & 0 & \mathbf{0}_{p\times(q-p)} \\
\cdots & \cdots & \cdots & \cdots & \cdots & \cdots & \cdots & \cdots & \\
0 & 0 & \cdots & \rho_p & 0 & 0 & \cdots & 1 & \\
\hdashline
 & \mathbf{0}_{(q-p)\times p} & & & & \mathbf{0}_{(q-p)\times p} & & & \begin{matrix} 1 & 0 & \cdots & 0 \\ 0 & 1 & \cdots & 0 \\ \cdots & \cdots & \cdots & \cdots \\ 0 & 0 & \cdots & 1 \end{matrix}
\end{array}\right), \tag{21.2}
$$

(The subscripts of the null submatrices show their dimensions.)

The variates $\xi_1, \ldots, \xi_p$ and $\eta_1, \ldots, \eta_q$ are known as canonical variates. They have the following properties as shown by their correlation matrix $\mathbf{P}$.

1. All the ξ's are uncorrelated with one another.
2. All the η's are uncorrelated with one another.
3. The pair of canonical variates ξ_i and η_i have correlation ρ_i $(i = 1, \ldots, p)$ but all other correlations between the ξ's and η's are zero. The ρ's are known as canonical correlations.

It is seen that by using the canonical variates the dependencies between the two original sets of data are reduced to their simplest possible form. The within-set correlations are all zero. The between-set correlations have been maximized (as shown below) between p pairs of variates, one from each set, and have been reduced to zero between all other pairs.

2. The Derivation of Canonical Variates

Consider the first pair of canonical variates ξ_1 and η_1. In what follows we shall, for clarity, omit the subscript 1 and write simply

$$\xi_1 = \xi = a_1 x_1 + a_2 x_2 + \cdots + a_p x_p = \mathbf{a}'\mathbf{x}$$

and

$$\eta_1 = \eta = b_1 y_{p+1} + b_2 y_{p+2} + \cdots + b_q y_{p+q} = \mathbf{b}'\mathbf{y}.$$

We now find the correlation between ξ and η, namely,

$$\text{cov}(\xi, \eta)/\{\text{var}(\xi)\text{var}(\eta)\}^{1/2}.$$

First it is necessary to obtain expressions for the variances and covariance. Clearly

$$\begin{aligned}\operatorname{var}(\xi) &= \operatorname{var}(a_1 x_1 + a_2 x_2 + \cdots + a_p x_p)\\ &= \sum_{i=1}^{p} a_i^2 \operatorname{var}(x_i) + \sum_{i \neq j} a_i a_j \operatorname{cov}(x_i, x_j).\end{aligned}$$

In terms of the symbols in (21.1) we then have

$$\begin{aligned}\operatorname{var}(\xi) &= \sum_{i=1}^{p} \sum_{j=1}^{p} a_i a_j \sigma_{ij}\\ &= \mathbf{a}'\mathbf{\Sigma}_{11}\mathbf{a}.\end{aligned}$$

Similarly

$$\operatorname{var}(\eta) = \sum_{i=1}^{q} \sum_{j=1}^{q} b_i b_j \sigma_{p+i,\, p+j} = \mathbf{b}'\mathbf{\Sigma}_{22}\mathbf{b}$$

and

$$\operatorname{cov}(\xi, \eta) = \sum_{i=1}^{p} \sum_{j=1}^{q} a_i b_j \sigma_{i,\, p+j} = \mathbf{a}'\mathbf{\Sigma}_{12}\mathbf{b} = \mathbf{b}'\mathbf{\Sigma}_{12}'\mathbf{a}.$$

The correlation between ξ and η is therefore

$$\rho(\mathbf{a}, \mathbf{b}) = \frac{\mathbf{a}'\mathbf{\Sigma}_{12}\mathbf{b}}{\{\mathbf{a}'\mathbf{\Sigma}_{11}\mathbf{a} \cdot \mathbf{b}'\mathbf{\Sigma}_{22}\mathbf{b}\}^{1/2}}.$$

We have written $\rho(\mathbf{a}, \mathbf{b})$ for the correlation to emphasize that it is a function of $\mathbf{a}$ and $\mathbf{b}$. We now determine what the elements of $\mathbf{a}$ and $\mathbf{b}$ must be for ξ and η to have maximum correlation (positive or negative) and for them to have arbitrary preassigned variances, v_ξ and v_η, say (this is equivalent to choosing scales for ξ and η). In other words, we wish to maximize $|\rho(\mathbf{a}, \mathbf{b})|$ subject to the constraints

$$\mathbf{a}'\mathbf{\Sigma}_{11}\mathbf{a} = v_\xi \quad \text{and} \quad \mathbf{b}'\mathbf{\Sigma}_{22}\mathbf{b} = v_\eta.$$

Write

$$f(\mathbf{a}, \mathbf{b}) = \mathbf{a}'\mathbf{\Sigma}_{12}\mathbf{b} - \frac{\lambda}{2}(\mathbf{a}'\mathbf{\Sigma}_{11}\mathbf{a} - v_\xi) - \frac{\mu}{2}(\mathbf{b}'\mathbf{\Sigma}_{22}\mathbf{b} - v_\eta),$$

where $\lambda/2$ and $\mu/2$ are undetermined (Lagrange) multipliers. Differentiating $f(\mathbf{a}, \mathbf{b})$ with respect to the elements of $\mathbf{a}$ gives

$$\frac{\partial f(\mathbf{a}, \mathbf{b})}{\partial a_i} = \sum_{j=1}^{q} b_j \sigma_{i,\, p+j} - \frac{\lambda}{2} \cdot 2 \sum_{j=1}^{p} a_j \sigma_{ij}, \qquad (i = 1, \ldots, p).$$

Similarly, differentiating with respect to the elements of **b** gives

$$\frac{\partial f(\mathbf{a}, \mathbf{b})}{\partial b_j} = \sum_{i=1}^{p} a_i \sigma_{i,\,p+j} - \frac{\mu}{2} \cdot 2 \sum_{i=1}^{q} b_i \sigma_{p+i,\,p+j}, \qquad (j = 1, \ldots, q).$$

Therefore

$$\frac{\partial f(\mathbf{a}, \mathbf{b})}{\partial \mathbf{a}} = \begin{pmatrix} \dfrac{\partial f}{\partial a_1} \\ \dfrac{\partial f}{\partial a_2} \\ \vdots \\ \dfrac{\partial f}{\partial a_p} \end{pmatrix} = \mathbf{\Sigma}_{12}\,\mathbf{b} - \lambda \mathbf{\Sigma}_{11}\mathbf{a}$$

and likewise

$$\frac{\partial f(\mathbf{a}, \mathbf{b})}{\partial \mathbf{b}} = \mathbf{\Sigma}_{12}'\,\mathbf{a} - \mu \mathbf{\Sigma}_{22}\,\mathbf{b}.$$

To find the values of the a's and b's that will maximize $|\rho(\mathbf{a}, \mathbf{b})|$ we now set these partial derivatives equal to zero and obtain the equations

$$\mathbf{\Sigma}_{12}\,\mathbf{b} - \lambda \mathbf{\Sigma}_{11}\mathbf{a} = \mathbf{0} \tag{21.3}$$

(both sides are p-element column vectors) and

$$\mathbf{\Sigma}_{12}'\,\mathbf{a} - \mu \mathbf{\Sigma}_{22}\,\mathbf{b} = \mathbf{0} \tag{21.4}$$

(both sides are q-element column vectors). Premultiply (21.3) by $\mathbf{a}'$ and (21.4) by $\mathbf{b}'$ to give the equations (in scalars)

$$\begin{aligned} \mathbf{a}'\mathbf{\Sigma}_{12}\,\mathbf{b} - \lambda \mathbf{a}'\mathbf{\Sigma}_{11}\mathbf{a} &= 0, \\ \mathbf{b}'\mathbf{\Sigma}_{12}'\mathbf{a} - \mu \mathbf{b}'\mathbf{\Sigma}_{22}\,\mathbf{b} &= 0. \end{aligned} \tag{21.5}$$

If we now scale the variates so that each has unit variance, that is, so that $\mathbf{a}'\mathbf{\Sigma}_{11}\mathbf{a} = \mathbf{b}'\mathbf{\Sigma}_{22}\,\mathbf{b} = 1$, we see that $\rho(\mathbf{a}, \mathbf{b})$ becomes $\mathbf{a}'\mathbf{\Sigma}_{12}\,\mathbf{b} = \mathbf{b}'\mathbf{\Sigma}_{12}'\,\mathbf{a}$ and, from (21.5), that

$$\rho(\mathbf{a}, \mathbf{b}) = \lambda = \mu.$$

Now premultiply the $p \times 1$ matrix equation (21.3) by the $q \times p$ matrix $\mathbf{\Sigma}_{12}'\mathbf{\Sigma}_{11}^{-1}$ to give the $q \times 1$ equation

$$\mathbf{\Sigma}_{12}'\mathbf{\Sigma}_{11}^{-1}\mathbf{\Sigma}_{12}\,\mathbf{b} - \lambda \mathbf{\Sigma}_{12}'\,\mathbf{a} = \mathbf{0}. \tag{21.6}$$

From (21.4) and noting that $\lambda = \mu$ we have

$$\mathbf{\Sigma}_{12}'\,\mathbf{a} = \lambda \mathbf{\Sigma}_{22}\,\mathbf{b}. \tag{21.7}$$

Substituting from (21.7) into (21.6) gives

$$(\boldsymbol{\Sigma}_{12}' \boldsymbol{\Sigma}_{11}^{-1} \boldsymbol{\Sigma}_{12} - \lambda^2 \boldsymbol{\Sigma}_{22})\mathbf{b} = \mathbf{0}. \tag{21.8}$$

Similarly, (21.4) leads to the $p \times 1$ equation

$$(\boldsymbol{\Sigma}_{12} \boldsymbol{\Sigma}_{22}^{-1} \boldsymbol{\Sigma}_{12}' - \lambda^2 \boldsymbol{\Sigma}_{11})\mathbf{a} = \mathbf{0}. \tag{21.9}$$

For (21.8) and (21.9) to have nontrivial solutions their determinants must vanish; that is, it is necessary that

$$|\boldsymbol{\Sigma}_{12}' \boldsymbol{\Sigma}_{11}^{-1} \boldsymbol{\Sigma}_{12} - \lambda^2 \boldsymbol{\Sigma}_{22}| = 0 \tag{21.10}$$

and

$$|\boldsymbol{\Sigma}_{12} \boldsymbol{\Sigma}_{22}^{-1} \boldsymbol{\Sigma}_{12}' - \lambda^2 \boldsymbol{\Sigma}_{11}| = 0; \tag{21.11}$$

(21.10) is of degree q in λ^2 and (21.11) is of degree p in λ^2. It can be shown (e.g., see Kendall and Stuart, 1966) that the nonzero roots of the two equations, of which there are p at most, are identical; (recall that $p \leq q$). Thus (21.10) [or (21.11)] may be solved to yield the roots $\lambda_1^2, \lambda_2^2, \ldots, \lambda_p^2$. Let them be ranked in order of decreasing size so that $\lambda_1^2 > \lambda_2^2 > \cdots > \lambda_p^2$.

Substitution of the numerical value of λ_1^2 into (21.8) and (21.9) enables us to solve these equations giving the required values of the a's and b's. Then the first pair of canonical variates is given by

$$\xi = \sum_{i=1}^{p} a_i x_i, \qquad \eta = \sum_{j=1}^{q} b_j y_{p+j}.$$

The 2nd, ..., pth pairs of canonical variates are found similarly by substituting $\lambda_2^2, \ldots, \lambda_p^2$, respectively into (21.8) and (21.9).

It may be proved (e.g., see Morrison, 1967) that, when $i \neq j$, $\text{cov}(\xi_i, \eta_j) = 0$. As a result, the correlation matrix of the ξ's and η's is as shown in (21.2).

3. A Numerical Example of Canonical Variate Analysis

It is worthwhile to demonstrate the method with a deliberately simplified artificial numerical example.

Imagine that 10 sample units of vegetation are examined, on each of which two environmental factors are measured to give x_1 and x_2 and the quantities of two species are recorded to give y_3 and y_4. Let the results be as shown in the table that follows. The left half shows the raw data and the right half the canonical variate values for each unit, which we now derive. (Note that the collection of 10 units is treated as a population, not as a sample, so that we

Raw Data				Canonical Variate Values			
Environmental Variates		Species Quantities		First Pair		Second Pair	
x_1	x_2	y_3	y_4	ξ_1	η_1	ξ_2	η_2
3	2	17	19	8.22	34.02	−3.88	0.11
4	4	14	18	14.44	30.13	−9.76	−2.00
8	3	18	16	15.83	32.34	−2.32	3.78
12	6	10	18	27.66	26.13	−8.64	−6.00
14	6	12	13	29.66	23.65	−6.64	0.44
16	7	10	12	34.27	20.75	−8.08	−0.67
21	8	7	11	41.88	16.86	−6.52	−2.78
24	8	9	8	44.88	16.17	−3.52	1.89
28	9	3	5	51.49	7.48	−2.96	−1.45
30	11	3	2	58.71	4.79	−7.84	1.22

put, for instance, $\text{var}(x_1) = (\frac{1}{10})\sum_{i=1}^{10}(x_{1i} - \bar{x}_1)^2$.) The covariance matrix of the x's and y's is

$$\begin{pmatrix} \mathbf{\Sigma}_{11} & \mathbf{\Sigma}_{12} \\ \mathbf{\Sigma}'_{12} & \mathbf{\Sigma}_{22} \end{pmatrix} = \begin{pmatrix} 82.60 & 23.20 & -41.20 & -48.20 \\ & 7.04 & -12.52 & -13.38 \\ & & 24.01 & 23.24 \\ & & & 30.36 \end{pmatrix}.$$

Notice that x_1 and x_2 are positively correlated, as are y_3 and y_4; but all the between-set correlations (x_1 and y_3; x_1 and y_4; x_2 and y_3; x_2 and y_4) are negative.

Next, evaluate $\mathbf{\Sigma}_{11}^{-1}$ and $\mathbf{\Sigma}_{22}^{-1}$:

$$\mathbf{\Sigma}_{11}^{-1} = \frac{1}{|\mathbf{\Sigma}_{11}|}\begin{pmatrix} \sigma_{22} & -\sigma_{21} \\ -\sigma_{12} & \sigma_{11} \end{pmatrix} = \frac{1}{43{\cdot}264}\begin{pmatrix} 7.04 & -23.20 \\ -23.20 & 82.60 \end{pmatrix}.$$

Likewise,

$$\mathbf{\Sigma}_{22}^{-1} = \frac{1}{188{\cdot}846}\begin{pmatrix} 30.36 & -23.24 \\ -23.24 & 24.01 \end{pmatrix}.$$

Evaluating $\mathbf{\Sigma}_{12}\mathbf{\Sigma}_{22}^{-1}\mathbf{\Sigma}'_{12}$ and substituting in (21.11) gives the determinantal equation

$$\begin{vmatrix} 79.5009 - 82.60\lambda^2 & 22.8184 - 23.20\lambda^2 \\ 22.8184 - 23.20\lambda^2 & 6.7309 - 7.04\lambda^2 \end{vmatrix} = 0$$

or

$$43.264\lambda^4 - 56.88492\lambda^2 + 14.43323 = 0$$

with roots

$$\lambda_1{}^2 = 0.9714 \quad \text{and} \quad \lambda_2{}^2 = 0.3434.$$

(It will be found that (21.10) has the same roots.)

Using the larger root λ_1^2 and writing a_{11}, a_{12}, b_{11}, and b_{12} for the coefficients of the first pair of canonical variates, (21.8) becomes

$$\begin{pmatrix} -0.73674 & 0.28192 \\ 0.28192 & -0.10776 \end{pmatrix} \begin{pmatrix} a_{11} \\ a_{12} \end{pmatrix} = \mathbf{0}.$$

One of the a's must be chosen arbitrarily. It is convenient to put $a_{11} = 1$. Then $a_{12} = 2.61$. Similarly (21.9) becomes

$$\begin{pmatrix} -1.05706 & 1.18029 \\ 1.18029 & -1.31764 \end{pmatrix} \begin{pmatrix} b_{11} \\ b_{12} \end{pmatrix} = \mathbf{0},$$

with a solution $b_{11} = 1$ and $b_{12} = 0.90$. The first pair of canonical variates is therefore given by

$$\begin{aligned} \xi_1 &= a_{11}x_1 + a_{12}x_2 = x_1 + 2.61x_2, \\ \eta_1 &= b_{11}y_3 + b_{12}y_4 = y_3 + 0.90y_4. \end{aligned} \tag{21.12}$$

The second pair of canonical variates ξ_2 and η_2 is obtained by substituting the smaller root $\lambda_2^2 = 0.3434$ in (21.8) and (21.9). It is found that

$$\begin{aligned} \xi_2 &= a_{21}x_1 + a_{22}x_2 = x_1 - 3.44x_2, \\ \eta_2 &= b_{21}y_3 + b_{22}y_4 = y_3 - 0.89y_4. \end{aligned}$$

Each of the original sample points (x_1, x_2, y_3, y_4) may now be expressed in terms of the canonical variates $(\xi_1, \xi_2, \eta_1, \eta_2)$. Their covariance matrix is as follows:

$$\begin{pmatrix} \text{var}(\xi_1) & \text{cov}(\xi_1, \xi_2) & \text{cov}(\xi_1, \eta_1) & \text{cov}(\xi_1, \eta_2) \\ & \text{var}(\xi_2) & \text{cov}(\xi_2, \eta_1) & \text{cov}(\xi_2, \eta_2) \\ & & \text{var}(\eta_1) & \text{cov}(\eta_1, \eta_2) \\ & & & \text{var}(\eta_2) \end{pmatrix}$$

$$= \begin{pmatrix} 251.661 & 0 & -148.354 & 0 \\ & 6.293 & 0 & 3.800 \\ & & 90.030 & 0 \\ & & & 6.683 \end{pmatrix}.$$

The correlation matrix is

$$\begin{pmatrix} 1 & 0 & -0.9856 & 0 \\ & 1 & 0 & 0.5860 \\ & & 1 & 0 \\ & & & 1 \end{pmatrix}.$$

Notice that the correlation between ξ_1 and η_1 is

$$\rho_{\xi_1\eta_1} = -0.9856, \quad \text{whence} \quad \rho^2_{\xi_1\eta_1} = \lambda_1^2 = 0.9714.$$

Likewise,

$$\rho_{\xi_2\eta_2} = 0.5860, \quad \text{whence} \quad \rho^2_{\xi_2\eta_2} = \lambda_2{}^2 = 0.3434.$$

These are the *only* nonzero correlations and the purpose of the analysis has been achieved. Attempts to interpret the result are not worthwhile, since the data have been invented for illustrative purposes.

4. Possible Ecological Applications of Canonical Analysis

In the foregoing discussion of canonical variate analysis it has been assumed throughout that the observed variate values were to be treated as constituting a *population*, not a sample. The problem of testing the significance of the canonical correlations did not therefore arise. For an account of the sampling theory and for ways of testing for the independence between two sets of variates see, for example, Anderson (1958) and Morrison (1967). However, in studies of ecological communities we are generally on safer ground if we treat the data at hand as material to be studied for its own sake. Attempts to make inferences about putative "parent populations" often lead to a welter of statistical difficulties and to the proliferation of unconvincing significance tests.

Canonical variate analysis seems to have been little used by ecologists. Buzas (1967) used it as a method for comparing the populations of foraminifera in sediments collected from bays and from the open ocean on the continental shelf in the Gulf of Mexico. This method has so far been most extensively used in educational and psychological research. Thus educationists may want to study the correlation between the grades and marks achieved by students, first at school and later in a university (Barnett and Lewis, 1963), or between the marks gained by children in a set of tests of their reading ability and another of their arithmetic ability [Hotelling (1936) quoted in Kendall and Stuart, 1966]. An example of an application in psychology is given by Morrison (1967); sets of test marks designed to measure adult intelligence were correlated with another set of variates measuring the subjects' accumulated experience. An application in genetics is described by Anderson (1958) and DeGroot and Li (1966). The first set of variates was obtained by taking measurements of head size on a sample of adult men; and the second set consisted of the same measurements obtained from each man's younger brother.

Ecologists are often confronted with data that consist of two (or more) different sets of observations. An obvious example, already discussed, is when one set lists the quantities of many different species in sample areas of vegetation and the other set is of measurements on environmental variables in the same areas. Another interesting possibility would be to take as sets of variates the numbers of individuals of species at two different trophic levels in samples from an animal community. Numerous possibilities suggest themselves and canonical analysis as an exploratory method deserves more attention from ecologists.

Bibliography

Anderson, T. W. (1958). *An Introduction to Multivariate Statistical Analysis*. Wiley, New York.

Anscombe, F. J. (1950). Sampling theory of the negative binomial and logarithmic series distributions. *Biometrika* **37**: 358–382.

Archibald, E. E. A. (1948). Plant populations. I. A new application of Neyman's contagious distribution. *Ann. Bot. Lond.* N.S. **12**: 221–235.

Bailey, N. T. J. (1964). *The Elements of Stochastic Processes with Applications to the Natural Sciences*. Wiley, New York.

Barnard, G. A. (1947). Significance tests for 2 × 2 tables. *Biometrika* **34**: 123–138.

Barnett, V. D. (1962). The Monte Carlo solution of a competing species problem *Biometrics* **18**: 76–103.

Barnett, V. D., and T. Lewis (1963). A study of the relation between G.C.E. and degree results. *J. Roy. Statist. Soc.* **A126**: 187–226.

Bartlett, M. S. (1957). On theoretical models for competitive and predatory biological systems. *Biometrika* **44**: 27–42.

Bartlett, M. S. (1960). *Stochastic Population Models in Ecology and Epidemiology*. Methuen, London.

Barton, D. E., and F. N. David (1959). The dispersion of a number of species. *J. Roy. Statist. Soc.* **B21**: 190–194.

Basharin, G. P. (1959). On a statistical estimate for the entropy of a sequence of independent random variables. *Theory Probab., Applic.* **4**: 333–336.

Bennett, B. M., and C. Horst (1966). *Supplement to Tables for Testing Significance in a 2 × 2 Contingency Table*. Cambridge University Press, Cambridge.

Birch, L. C. (1948). The intrinsic rate of natural increase of an insect population. *J. Anim. Ecol.* **17**: 15–26.

Bliss, C. I. (1965). An analysis of some insect trap records. In "Classical and Contagious Discrete Distributions" (G. P. Patil, Ed.), Statistical Publishing Society, Calcutta.

Bliss, C. I., and R. A. Fisher (1953). Fitting the negative binomial distribution to biological data and a note on the efficient fitting of the negative binomial. *Biometrics* **9**: 176–200.

Bray, J. R. (1956). A study of mutual occurrence of plant species. *Ecology* **37**: 21–28.

Brian, M. V. (1953). Species frequencies in random samples from animal populations. *J. Anim. Ecol.* **22**: 57–64.

Brillouin, L. (1962). *Science and Information Theory*. 2nd ed. Academic, New York.

Broadbent, S. R., and D. G. Kendall (1953). The random walk of *Trichostrongylus retortaeformis*. *Biometrics* **9**: 460–466.

Buzas, M. A. (1967). An application of canonical analysis as a method for comparing faunal areas. *J. Anim. Ecol.* **36**: 563–577.

Caughley, G. (1966). Mortality patterns in mammals. *Ecology* **47**: 906–918.

Caughley, G. (1967). Parameters for seasonally breeding populations. *Ecology* **48**: 834–839.

Chiang, C. L. (1954). Competition and other interactions between species. In *Statistics and Mathematics in Biology* (O. Kempthorne et al., Eds.). Iowa State College Press, Ames.

Chiang, C. L. (1960a). A stochastic study of the life table and its applications: I. Probability distributions of the biometric functions. *Biometrics* **16**: 618–635.

Chiang, C. L. (1960b). A stochastic study of the life table and its applications: II. Sample variance of the observed expectation of life and other biometric functions. *Human Biology* **32**: 221–238.

Clark, P. J., and F. C. Evans (1954). Distance to nearest neighbor as a measure of spatial relationships in populations. *Ecology* **35**: 445–453.

Clark, P. J., and F. C. Evans (1955). On some aspects of spatial pattern in biological populations. *Science* **121**: 397–398.

Cochran, W. G. (1954). Some methods for strengthening the common χ^2-tests. *Biometrics* **10**: 417–451.

Cohen, J. E. (1966). *A Model of Simple Competition*. Harvard University Press, Cambridge, Mass.

Cohen, J. E. (1968). Alternate derivations of a species-abundance relation. *Amer. Natur.* **102**: 165–172.

Cole, L. C. (1949). The measurement of interspecific association. *Ecology* **30**: 411–424.

Cooper, C. F. (1961). Pattern in Ponderosa pine forests. *Ecology* **42**: 493–499.

Dagnelie, P. (1966). A propos des différentes méthodes de classification numerique. *Revue Statist. Appliq.* **14**: 55–75.

David, F. N., and P. G. Moore, (1954). Notes on contagious distributions in plant populations. *Ann. Bot. Lond.* N.S. **18**: 47–53.

DeBach, P. (1966). The competitive displacement and coexistence principles. *Ann. Rev. Ent.* **11**: 183–212.

Deevey, E. S. (1947). Life tables for natural populations of animals. *Quart. Rev. Biol.* **22**: 283–314.

DeGroot, M. H., and C. C. Li (1966). Correlations between similar sets of measurements. *Biometrics* **22**: 781–790.

Dice, L. R. (1945). Measures of the amount of ecologic association between species. *Ecology* **26**: 297–302.

Dublin, L. I., and Lotka, A. J. (1925). On the true rate of natural increase. *J. Amer. Stat. Ass.* **20**: 305–339.

Edwards, A. W. F., and L. L. Cavalli-Sforza (1965). A method for cluster analysis. *Biometrics* **21**: 362–375.

Feller, W. (1943). On a general class of contagious distributions. *Ann. Math. Statist.* **14**: 389–400.

Feller, W. (1966). *An Introduction to Probability Theory and Its Applications*, Vol. II. Wiley, New York.

Feller, W. (1968). *An Introduction to Probability Theory and Its Applications*, Vol. I, 3rd ed. Wiley, New York.

Finney, D. J., R. Latscha, B. M. Bennett, and P. Hsu (1963). *Tables for Testing Significance in a 2 × 2 Contingency Table.* Cambridge University Press, Cambridge.

Fisher, R. A., A. S. Corbet, and C. B. Williams (1943). The relation between the number of species and the number of individuals in a random sample of an animal population. *J. Anim. Ecol.* **12**: 42–58.

Frank, P. W. (1968). Life histories and community stability. *Ecology* **49**: 355–357.

Gause, G. F. (1934). *The Struggle for Existence.* Hafner, New York.

Gittins, R. (1965). Multivariate approaches to a limestone grassland community: I. A stand ordination. *J. Ecol.* **53**: 385–401.

Good, I. J. (1950). *Probability and the Weighing of Evidence.* Griffin, London.

Good, I. J. (1953). The population frequencies of species and the estimation of population parameters. *Biometrika* **40**: 237–264.

Goodall, D. W. (1954). Objective methods for the classification of vegetation: III. An essay in the use of factor analysis. *Aust. J. Bot.* **2**: 304–324.

Goodall, D. W. (1966). Hypothesis testing in classification. *Nature* **211**: 329–330.

Gower, J. C. (1966). Some distance properties of latent root and vector methods used in multivariate analysis. *Biometrika* **53**: 325–338.

Gower, J. C. (1967a). A comparison of some methods of cluster analysis. *Biometrics* **23**: 623–637.

Gower, J. C. (1967b). Multivariate analysis and multidimensional geometry. *The Statistician* **17**: 13–28.

Greig-Smith, P. (1952). The use of random and contiguous quadrats in the study of the structure of plant communities. *Ann. Bot. Lond.* N.S. **16**: 293–316.

Greig-Smith, P. (1964). *Quantitative Plant Ecology.* 2nd ed. Butterworths, London.

Greig-Smith, P., M. P. Austin, and T. C. Whitmore (1967). The application of quantitative methods to vegetation survey: I. Association analysis and principal component ordination of rain forest. *J. Ecol.* **55**: 483–503.

Grundy, P. M. (1951). The expected frequencies in a sample of an animal population in which the abundances of species are log-normally distributed. Part I. *Biometrika* **38**: 427–434.

Hairston, N. G. (1967). Studies on the limitation of a natural population of *Paramecium aurelia*. *Ecology* **48**: 904–909.

Herdan, G. (1958). The mathematical relation between Greenberg's index of linguistic diversity and Yule's characteristic. *Biometrika* **45**: 268–270.

Hoel, P. (1954). *Introduction to Mathematical Statistics.* Wiley, New York.

Hopkins, B. (1955). The species area relations of plant communities. *J. Ecol.* **43**: 409–426.

Hopkins, B., and J. G. Skellam (1954). A new method for determining the type of distribution of plant individuals. *Ann. Bot. Lond.* N.S. **18**: 213–227.

Howard, R. A. (1960). *Dynamic Programming and Markov Processes.* Wiley, New York.

Kemeny, J. G., and J. L. Snell (1960). *Finite Markov Chains.* van Nostrand, Princeton, N.J.

Kendall, M. G. (1957). *A Course in Multivariate Analysis.* Griffin, London.

Kendall, M. G., and A. Stuart (1966). *The Advanced Theory of Statistics.* Vol. 3. Griffin, London.

Kendall, M. G., and A. Stuart, (1967). *The Advanced Theory of Statistics.* Vol. 2. 2nd ed. Griffin, London.

Kershaw, K. A. (1960). The detection of pattern and association. *J. Ecol.* **48**: 233–242.

Kershaw, K. A. (1961). Association and co-variance analysis of plant communities. *J. Ecol.* **49**: 643–654.

Khinchin, A. I. (1957). *Mathematical Foundations of Information Theory.* Dover, New York.

Kilburn, P. D. (1966). Analysis of the species area relation. *Ecology* **47**: 831–843.

Krishna Iyer, P. V. (1949). The first and second moments of some probability distributions arising from points on a lattice and their application. *Biometrika* **36**: 135–141.

Krishna Iyer, P. V. (1952). Factorial moments and cumulants of distributions arising in Markoff chains. *J. Ind. Soc. Agr. Stat.* **4**: 113–123.

Lawley, D. N., and A. E. Maxwell (1963). *Factor Analysis as a Statistical Method.* Butterworths, London.

Lefkovitch, L. P. (1965). The study of population growth in organisms grouped by stages. *Biometrics* **21**: 1–18.

Leigh, E. J. Jr. (1968). Review of K. E. F. Watt's "Ecology and Resource Management." *Science* **160**: 1326–1327.

Leslie, P. H. (1945). The use of matrices in certain population mathematics. *Biometrika* **33**: 183–212.

Leslie, P. H. (1948). Some further notes on the use of matrices in population mathematics. *Biometrika* **35**: 213–245.

Leslie, P. H. (1958). A stochastic model for studying the properties of certain biological systems by numerical methods. *Biometrika* **45**: 16–31.

Leslie, P. H. (1959). The properties of a certain lag type of population growth and the influence of an external random factor on a number of such populations. *Physiol. Zool.* **32**: 151–159.

Leslie, P. H., and J. C. Gower (1960). The properties of a stochastic model for the predator-prey type of interaction between two species. *Biometrika* **47**: 219–234.

Leslie, P. H., and T. Park (1949). The intrinsic rate of natural increase of *Tribolium castaneum* Herbst. *Ecology* **30**: 469–477.

Leslie, P. H., and R. M. Ranson (1940). The mortality, fertility and rate of natural increase of the vole (*Microtus agrestis*) as observed in the laboratory. *J. Anim. Ecol.* **9**: 27–52.

Lewis, E. G. (1942). On the generation and growth of a population. *Sankhyā* **6**: 93–96.

Lloyd, M. (1967). Mean crowding. *J. Anim. Ecol.* **36**: 1–30.

Lotka, A. J. (1925). *Elements of Physical Biology.* Williams & Wilkins, Baltimore.

MacArthur, R. H. (1957). On the relative abundance of bird species. *Proc. Nat. Acad. Sci. Wash.* **43**: 293–295.

Margalef, D. R. (1958). Information theory in ecology. *General Systems* **3**: 36–71.

Matérn, B. (1960). Spatial variation. Stochastic models and their application to some problems in forest surveys and other sampling investigations. *Medd. fran Statens Skogsforskningsinstitut.* **49**: 1–144.

McIntosh, R. P. (1967a). The continuum concept of vegetation. *Bot Rev.* **33**: 130–187.

McIntosh, R. P. (1967b). An index of diversity and the relation of certain concepts to diversity. *Ecology* **48**: 392–404.

Miller, G. A., and W. G. Madow (1954). On the maximum likelihood estimate of the Shannon-Wiener measure of information. AFCRC-TR-54-75, Air Force Cambridge Research Center, Bolling Air Force Base, Washington, D.C.

Moore, P. G. (1954). Spacing in plant populations. *Ecology* **35**: 222–227.

Morisita, M. (1954). Estimation of population density by spacing method. *Mem. Fac. Sci. Kyushu U. Series E.* (*Biol.*) **1**: 187–197.

Morisita, M. (1959). Measuring the dispersion of individuals and analysis of the distributional patterns. *Mem. Fac. Sci. Kyushu U. Series E.* (*Biology*) **2**: 215–235.

Morrison, D. F. (1967). *Multivariate Statistical Methods.* Wiley, New York.

Mountford, M. D. (1961). On E. C. Pielou's index of non-randomness. *J. Ecol.* **49**: 271–275.

Nelson, W. C. and H. A. David (1967). The logarithmic distribution: a review. *Virginia Journal of Sci.* **18**: 95-102.

Neyman, J., T. Park, and E. L. Scott (1956). Struggle for existence. The *Tribolium* model. *Proc. 3rd Berkeley Symposium on Math. Stat. and Prob.* **3**: 41–79.

Odum, E. P. (1959). *Fundamentals of Ecology*, 2nd ed. Saunders, Philadelphia.

Orloci, L. (1966). Geometric models in ecology: I. The theory and application of some ordination methods. *J. Ecol.* **54**: 193–215.

Orloci, L. (1967a). An agglomerative method for classification of plant communities. *J. Ecol.* **55**: 193–205.

Orloci, L. (1967b). Data centering: a review and evaluation with reference to component analysis. *System. Zool.* **16**: 208–212.

Park, T. (1954). Experimental studies on interspecies competition: II. Temperature, humidity and competition in two species of *Tribolium. Physiol. Zool.* **27**: 177–238.

Pearson, E. S. (1947). The choice of statistical tests illustrated on the interpretation of data classed in a 2×2 table. *Biometrika* **34**: 139–167.

Phillips, M. E. (1953). Studies in the quantitative morphology and ecology of *Eriophorum angustifolium* Roth.: I. The rhizome system. *J. Ecol.* **41**: 295–318.

Pielou, E. C. (1959). The use of point-to-plant distances in the study of the pattern of plant populations. *J. Ecol.* **47**: 607–613.

Pielou, E. C. (1961). Segregation and symmetry in two-species populations as studied by nearest neighbor relations. *J. Ecol.* **49**: 255–269.

Pielou, E. C. (1962a). The use of plant-to-neighbor distances for the detection of competition. *J. Ecol.* **50**: 357–367.

Pielou, E. C. (1962b). Runs of one species with respect to another in transects through plant populations. *Biometrics* **18**: 579–593.

Pielou, E. C. (1963a). The distribution of diseased trees with respect to healthy ones in a patchily infected forest. *Biometrics* **19**: 450–459.

Pielou, E. C. (1963b). Runs of healthy and diseased trees in transects through an infected forest. *Biometrics* **19**: 603–614.

Pielou, E. C. (1964). The spatial pattern of two-phase patchworks of vegetation. *Biometrics* **20**: 156–167.

Pielou, E. C. (1965a). The concept of randomness in the patterns of mosaics. *Biometrics.* **21**: 908–920.

Pielou, E. C. (1965b). The concept of segregation pattern in ecology: some discrete distributions applicable to the run lengths of plants in narrow transects. In "Classical and Contagious Discrete Distributions" (G. P. Patil, Ed.). Statistical Publishing Society, Calcutta.

Pielou, E. C. (1966a). Species-diversity and pattern-diversity in the study of ecological succession. *J. Theoret. Biol.* **10**: 370–383.

Pielou, E. C. (1966b). Shannon's formula as a measure of specific diversity: its use and misuse. *Amer. Natur.* **100**: 463–465.

Pielou, E. C. (1966c). The measurement of diversity in different types of biological collections. *J. Theoret. Biol.* **13**: 131–144.

Pielou, E. C. (1967a). The use of information theory in the study of the diversity of biological populations. *Proc. 5th Berkeley Symposium on Math. Stat. and Prob.* **4**: 163–177.

Pielou, E. C. (1967b). A test for random mingling of the phases of a mosaic. *Biometrics* **23**: 657–670.

Pielou, E. C., and R. E. Foster (1962). A test to compare the incidence of disease in isolated and crowded trees. *Can. J. Botany* **40**: 1176–1179.

Pielou, D. P., and W. G. Matthewman (1966). The fauna of *Fomes fomentarius* (Linnaeus ex Fries) Kickx. growing on dead birch in Gatineau Park, Quebec. *Can. Ent.* **98**: 1308–1312.

Pielou, D. P., and E. C. Pielou (1967). The detection of different degrees of coexistence. *J. Theoret. Biol.* **16**: 427–437.

Pielou, D. P., and E. C. Pielou (1968). Association among species of infrequent occurrence: The insect and spider fauna of *Polyporus betulinus* (Bulliard) Fries. *J. Theoret. Biol.* **21**: 202–216.

Pollard, J. H. (1966). On the use of the direct matrix product in analysing certain stochastic population models. *Biometrika* **53**: 397–415.

Preston, F. W. (1948). The commonness and rarity of species. *Ecology* **29**: 254–283.

Shannon, C. E., and W. Weaver (1949). *The Mathematical Theory of Communication.* University of Illinois Press, Urbana.

Simpson, E. H. (1949). Measurement of diversity, *Nature* **163**: 688.

Skellam, J. G. (1951). Random dispersal in theoretical populations. *Biometrika* **38**: 196–218.

Skellam, J. G. (1952). Studies in statistical ecology: I. Spatial pattern. *Biometrika* **39**: 346–362.

Slobodkin, L. B. (1961). *Growth and Regulation in Animal Populations.* Holt, Rinehart and Winston, New York.

Smith, F. E. (1963). Population dynamics in *Daphnia magna. Ecology* **44**: 651–663.

Sokal, R. R., and P. H. A. Sneath (1963). *Principles of Numerical Taxonomy.* Freeman, San Francisco.

Swed, F. S., and C. Eisenhart (1943). Tables for testing randomness of grouping in a sequence of alternatives. *Ann. Math. Statist.* **14**: 66–85.

Switzer, P. (1965). A random set process in the plane with a Markovian property. *Ann. Math. Statist.* **36**: 1859–1863.

Switzer, P. (1967). Reconstructing patterns from sample data. *Ann. Math. Statist.* **38**: 138–154.

Taylor, L. R. (1961). Aggregation, variance and the mean. *Nature* **189**: 732–735.

Thompson, H. R. (1956). Distribution of distance to *n*th neighbor in a population of randomly distributed individuals. *Ecology* **37**: 391–394.

Waters, W. E. (1959). A quantitative measure of aggregation in insects. *J. Econ. Ent,* **52**: 1180–1184.

Watson, G. N. (1944). *Theory of Bessel Functions.* Cambridge University Press, Cambridge.

Watt, A. S. (1947). Pattern and process in the plant community. *J. Ecol.* **35**: 1–22.

Watt, K. E. F. (1968). *Ecology and Resource Management.* McGraw-Hill, New York.

Whitworth, W. A. (1934). *Choice and Chance,* Steichert. New York.

Wilkinson, D. H. (1952). The random element in bird "navigation." *J. Experimental Biol.* **29**: 532–560.

Williams, C. B. (1964). *Patterns in the Balance of Nature.* Academic, New York.

Williams, E. J. (1961). The distribution of larvae of randomly moving insects. *Austr. J. Biol. Sci.* **14**: 598–604.

Williams, W. T., and J. M. Lambert (1959). Multivariate methods in plant ecology: I. Association-analysis in plant communities. *J. Ecol.* **47**: 83–101.

Williams, W. T., and J. M. Lambert (1960). Multivariate methods in plant ecology: II. The use of an electronic digital computer for association analysis. *J. Ecol.* **48**: 689–710.

Williams, W. T., and J. M. Lambert (1966). Multivariate methods in plant ecology: V. Similarity analyses and information-analysis. *J. Ecol.* **54**: 427–445.

Wylie, C. R. Jr. (1951). *Advanced Engineering Mathematics.* McGraw-Hill, New York.

Yule, G. U. (1912). On the methods of measuring association between two attributes. *J. Roy. Statist. Soc.* **75**: 579–642.

Yule, G. U. (1924). A mathematical theory of evolution based on the conclusions of Dr. J. C. Willis F.R.S. *Phil. Trans. Roy. Soc. Lond.* **B213**: 21–87.

Author Index

Subject Index